Statistical Analysis of Networks

Konstantin Avrachenkov and
Maximilien Dreveton

now
the essence of knowledge

Published, sold and distributed by:
now Publishers Inc.
PO Box 1024
Hanover, MA 02339
United States
Tel. +1-781-985-4510
www.nowpublishers.com
sales@nowpublishers.com

Outside North America:
now Publishers Inc.
PO Box 179
2600 AD Delft
The Netherlands
Tel. +31-6-51115274

ISBN: 978-1-63828-050-7
E-ISBN: 978-1-63828-051-4
DOI: 10.1561/9781638280514

Suggested citation: Konstantin Avrachenkov and Maximilien Dreveton. (2022). *Statistical Analysis of Networks*. Boston–Delft: Now Publishers

Dedicated to our mothers, Elena and Christine.

Table of Contents

Preface

This book is a general introduction to the statistical analysis of networks, and can serve both as a research monograph and as a textbook. Many fundamental modern tools and concepts needed for the analysis of networks are presented, such as network modeling, community detection, graph-based semi-supervised learning and sampling in networks. The description of these concepts is self-contained, with both theoretical justifications and applications provided for the presented algorithms.

Researchers, including postgraduate students, working in the area of network science, complex network analysis, or social network analysis, will find up-to-date statistical methods relevant to their research tasks. This book can also serve as textbook material for courses related to the statistical approach to the analysis of complex networks.

In general, the chapters are fairly independent and self-supporting, and the book could be used for course composition "à la carte". Nevertheless, Chapter 2 is needed to a certain degree for all parts of the book. It is also useful to read Chapter 4 before reading Chapters 5 and 6, but this is not absolutely necessary. Reading Chapter 3 can also be helpful before reading Chapters 5 and 7.

As prerequisites for reading our book, we expect basic knowledge in probability, linear algebra and elementary notions of graph theory. We have also added appendices describing some required notions from the above mentioned disciplines.

DOI: 10.1561/9781638280514.ch1

Chapter 1

Introduction

A *network* is a collection of objects interacting with each other. Networks are found in numerous scientific disciplines: atoms or interacting particles in statistical physics, protein interactions in molecular biology, social networks in sociology and the Internet web-graph in computer science, just to name a few. Several types of interactions exist. While binary interactions are the simplest (did Alice interact with Bob today?), weighted interactions (the number of interactions between Alice and Bob today) or temporal interactions (at what precise times did Alice and Bob interact?) provide additional valuable information.

Networks with binary interactions are conveniently represented by a *graph*. A graph G is a pair (V, E), where V is the set of objects (also called *nodes* or *vertices*), and E is the set of interacting node pairs (also called *edges* or *links*). This standard graph representation can be extended to weighted networks or temporal networks by considering weighted edges or temporal sequences of edges. In the first and second parts of the introductory section, we present several examples of real-world networks and describe unifying properties.

1.1 Examples of Networks

Let us present several examples of real-world networks. Although, for clarity of exposition, we categorise the networks by types, this classification is subjective, and a network could belong to two or more types.

Social networks

One of the first examples of social networks is the *Zachary karate club*, representing the friendships between the 74 members of a karate club (see Figure 1.1). During the two-year study (Zachary, 1977), the club members split into two groups after a feud occurred between the main instructor and the club's president. This dispute makes the dataset extremely popular in the network science community. We would like to answer the intriguing question: can one predict the resulting two groups based only on the friendship graph? This lays the ground for the problem of *community detection*, which we will discuss in detail in Chapter 4.

Data concerning social networks of real-life social relationships (acquaintances, interactions) are notoriously hard to gather. Indeed, questionnaires are physical and take time to analyse, making the collection from a large number of individuals difficult. Moreover, they are prone to human error and personal interpretation. Fortunately, it is much easier to gather examples of datasets in online social networks.

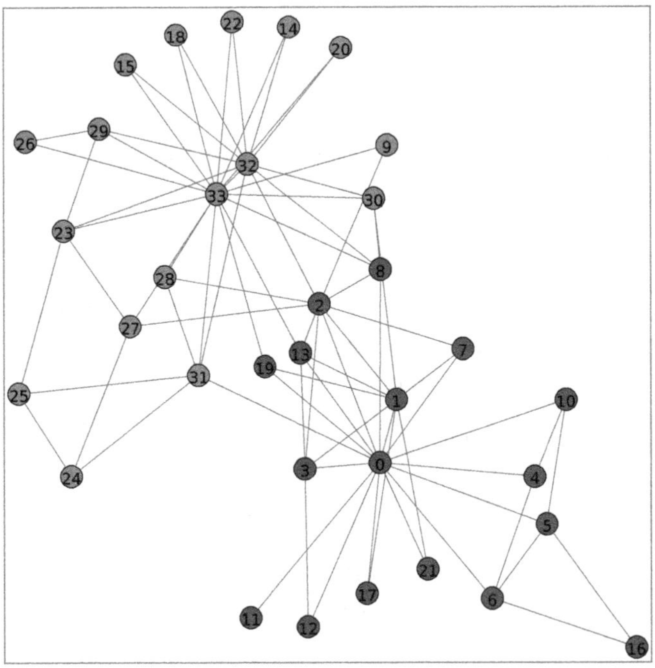

Figure 1.1. Karate club.

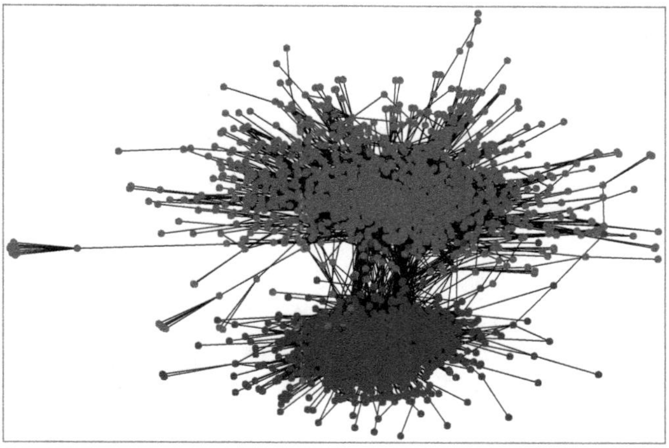

Figure 1.2. Two largest communities of the LiveJournal network.

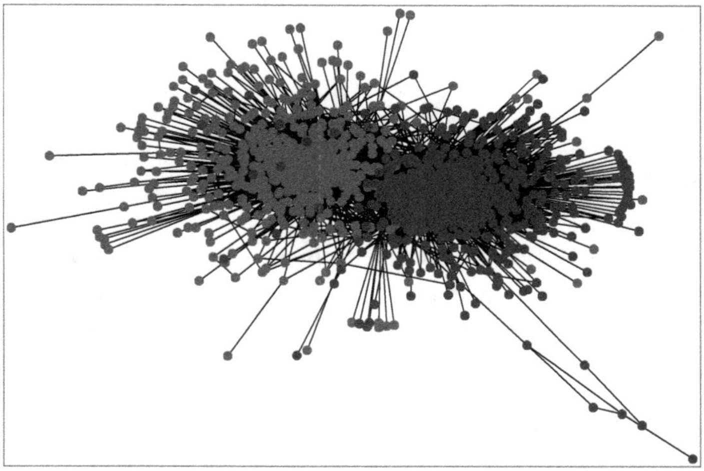

Figure 1.3. Political Blogs network.

Thus, it is not surprising that most examples of datasets of large social networks come from online social networks or web-blogs.

One such example is the *LiveJournal* dataset. LiveJournal is an online blogging community in which users can befriend each other. The users are also free to create groups which other users can join. These groups can be considered as ground-truth communities. Figure 1.2 shows the LiveJournal friendship network restricted to the two largest communities.

Adamic and Glance, 2005 studied the linking patterns of political bloggers during the U.S. Presidential Election of 2004. They considered 1494 blogs in total, 759 liberal and 735 conservative, and constructed the interactions by identifying whether one blog references another blog. As shown in Figure 1.3, the difference

Table 1.1. Dimensions of three data sets of interacting high school students: the number of students n, the number of classes K, and the number of snapshots T.

Year	n	K	T
2011	118	3	5609
2012	180	5	11273
2013	327	9	7375

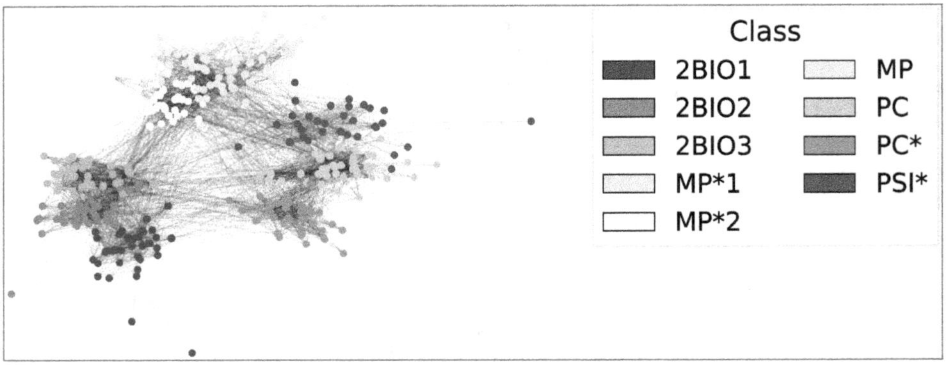

Figure 1.4. Time-aggregated network obtained from the high-school interaction network (year 2013).

between the liberal and conservative blogospheres is clear. Indeed, 90% of the interactions occur between blogs belonging to the same political community.

Some other prominent online social networks are *Twitter*, *Facebook* and *LinkedIn*.

Face-to-face interaction networks

The *high-school datasets* represent close proximity encounters between students in a French high school. Student-to-student interactions are recorded every 20 seconds through wearable sensors, and the experiments span several school days. The same experiment was performed in three consecutive years (Fournet and Barrat, 2014; Mastrandrea *et al.*, 2015), and the dimensions of each dataset are given in Table 1.1. We also plot in Figure 1.4 the weighted graph for the year 2013, where the weights correspond to the number of interactions recorded between two students. Finally, as each student belongs to one class, the question of recovering the classes based on the temporal interactions arises. We will study this dataset in more detail in Chapter 6.

We note that time aggregation can result in a loss of important information, which could otherwise be inferred from the dataset's temporal nature. For example,

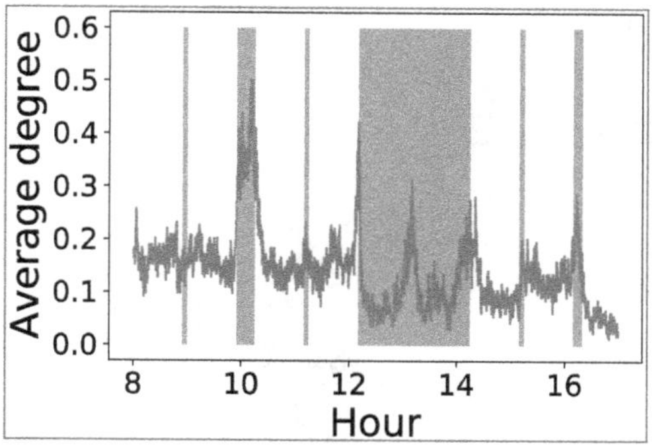

Figure 1.5. The average degree (the average number of interactions per student) over the course of a single day. The shaded regions correspond to the breaks between classes.

Figure 1.5 shows, per snapshot, the average number of interactions per student over a given day. The observed peaks correspond to the starting and ending times of the breaks between courses, since students leave and join the classrooms at these moments.

Communication networks

Communication networks constitute an important class, which includes various transportation networks (roads, airplane maps, etc.) as well as phone and messaging communications between individuals.

The *Enron email dataset*[1] contains approximately 500,000 emails from about 150 employees (mostly from the senior management team) of the Enron company (now bankrupt). Emails were recovered by the Federal Energy Regulatory Commission during the fraud investigation. This dataset was made public and has been used by many researchers for various information processing tasks, such as document classification or social network analysis (Carley and Skillicorn, 2005).

The *Copenhagen networks study dataset* (Sapiezynski *et al.*, 2019) records the interaction of 700 university students over 4 weeks, including close-proximity interactions, phone calls and Facebook friendships.

Information and collaboration networks

Co-authorship networks are constructed by connecting two authors if they have published a paper together. Since automated citation indexing is now common, large datasets of co-authorship networks are now available. Examples include the *DBLP* (Yang and Leskovec, 2015), *Citeseer*, *Cora*, *WebKB* (Getoor, 2005) and

1. Available at https://www.cs.cmu.edu/~enron/

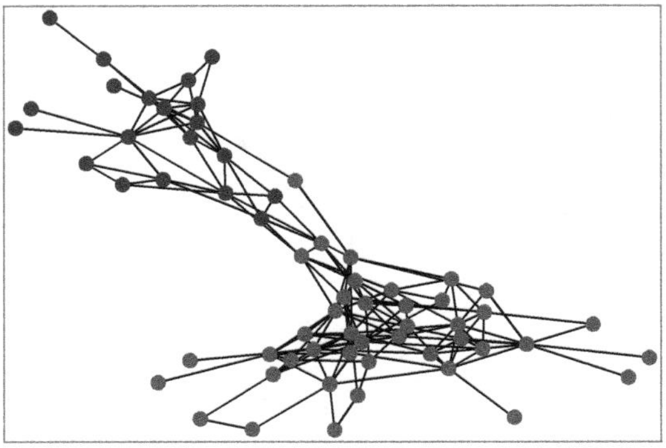

Figure 1.6. Dolphin network (Lusseau *et al.*, 2003). Colours show how the network split when a dolphin left the group.

PubMed (Namata *et al.*, 2012) datasets. This procedure can be extended to other domains. For example, using IMDB data one can produce a network of movie actors, where two actors are connected if they starred in a movie together.

Web-graph represents another example of information networks. It is constructed by linking webpage A to webpage B (usually with a directed link) if webpage A cites webpage B. Several Web-graph and Wikipedia networks are available from the NetSet database[2] and the Laboratory for Web Algorithmics (LAW).[3]

Biological networks

The class of biological networks includes protein interaction networks, food webs and animal social networks.

Let us present one example of an animal social network. The dolphin network (Lusseau *et al.*, 2003) is a social network of 62 dolphins, with edges representing social interactions. During the study, a dolphin left the group, which resulted in a split of the network into two communities (see Figure 1.6). The group later reunited when this mysterious dolphin returned home.

Geometrically defined network topologies

In machine learning tasks, data often come as a matrix

$$X = \left(x_1, \ldots, x_n\right) \in \mathbb{R}^{m \times n},$$

where n is the number of data points and m is the dimension of each data point (*e.g.*, the number of features). To perform data analysis with the help of a network,

2. https://netset.telecom-paris.fr/index.html

3. https://law.di.unimi.it/datasets.php

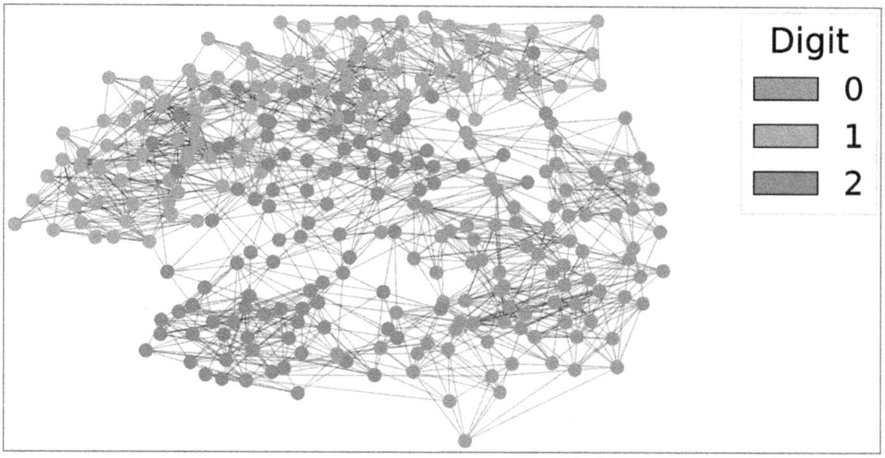

Figure 1.7. Network constructed from 300 pictures of digits 0, 1 and 2 taken from the MNIST database.

the topology and weights of the graph must be built from the data. A common way to define the weight of an edge connecting vertices i and j is by using a Gaussian kernel with thresholding

$$w_{ij} = \begin{cases} \exp\left(-\frac{\|x_i - x_j\|^2}{\tau^2}\right), & \text{if } \|x_i - x_j\|^2 \leq \kappa, \\ 0, & \text{otherwise,} \end{cases}$$

where τ and κ are tunable parameters and $\|\cdot\|$ is a distance between data points. In particular, the cutoff parameter κ prevents having a too dense network with many small-weight edges. Another common method is to connect each vertex to its K-nearest neighbours. We refer to (Grady and Polimeni, 2010, Chapter 4) and (Stankovic *et al.*, 2020) for the description of other methods for data similarity network construction.

The MNIST database (LeCun *et al.*, 1998) is a database of 70,000 handwritten digits commonly used as a benchmark in machine learning. Figure 1.7 presents a network built from 300 pictures of digits $0, 1, 2$ using the Gaussian kernel as a weight function. More precisely, we first compute a K-nearest neighbour graph ($K = 8$) with weights

$$w_{ij} = \begin{cases} \exp\left(-\frac{4\|x_i - x_j\|^2}{\tau_i}\right), & \text{if } x_j \text{ is among } K \text{ nearest neighbours of } x_i, \\ 0, & \text{otherwise,} \end{cases}$$

where τ_i represents the distance between x_i and its Kth-nearest neighbour. The weight matrix is finally symmetrised by replacing W with $\frac{1}{2}(W + W^T)$.

1.2 Unifying Properties of Complex Networks

1.2.1 What are the Properties Commonly Shared by Networks?

Many real-world complex networks share a number of basic properties.

Sparsity

The *degree* of node i, denoted d_i, is the number of edges incident to this node, or in other words, the number of nodes that are interacting with node i. Even if the number of nodes n in a network can be large, the average degree $\bar{d} = \frac{1}{n}\sum_{i=1}^{n} d_i$ is often small. For example, in Table 1.2, we see that the DBLP co-authorship network has 13,326 nodes, while the average degree $\bar{d}$ is just 5.1. This effect is even more evident in social networks such as *Facebook*. Even if the total number of users is huge and still growing, the number of friends of each user remains small (maybe even bounded). We say that a network is *sparse* if the average degree $\bar{d}$ is several orders of magnitude smaller than the number of nodes n.

Connectivity

A *connected component* of a binary undirected graph $G = (V, E)$ is a set U of nodes such that between two nodes $i, j \in U$ there exists a path linking i to j. Since two connected components are necessarily disjoint, the node set V can therefore be partitioned into a finite number of non-overlapping connected components $U_1, \ldots, U_p$. We say that the graph is *connected* if $p = 1$, and *disconnected* otherwise. Even though real-world networks might be disconnected, typically the relative size of the largest connected component is very large (for example, containing about 90% of the nodes), while the other components are much smaller (Newman, 2001a).

Small world

In a famous experiment, Milgram asked participants to mail a folder (containing several documents related to the study) to one of their acquaintances in an attempt to eventually reach an assigned target individual (Milgram, 1967). While in most cases the individuals failed (either by incapacity or lack of willingness), about 20% of the participants managed to send the documents to the assigned target.[4] Moreover, the mean number of intermediaries between starters and the target was 5.2. While Milgram's experiments were later criticized (Kleinfeld, 2002),

4. This is astonishing. Milgram's experiment was repeated using e-mail. Dodds *et al.*, 2003 asked 24,163 volunteers to start e-mail chains, aiming to reach 18 target persons in 13 countries. Only 384 (less than 1.6%) of those chains were completed!

Table 1.2. Basic characteristics of a selection of networks. The quantities are: the number of nodes n, the number of edges $|E|$, the average degree (the average number of neighbour nodes) $\bar{d}$, the average distance between two nodes δ, clustering coefficient cc (in parenthesis the clustering coefficient if the edges of the graph were drawn randomly), the exponent of the degree distribution α.

Network	n	$\|E\|$	$\bar{d}$	δ	cc	α
Political blogs	1222	16717	27.4	2.7	0.32 (0.07)	1.5
citeseer	2110	3720	3.5	9.3	0.17 (0.005)	2.7
cora	2485	5069	4.0	6.3	0.24 (0.005)	2.9
LiveJournal	2766	24138	17.5	3.9	0.41 (0.02)	2.1
wikischools	4403	100382	46	2.5	0.28 (0.03)	2.3
DBLP	13326	34281	5.1	6.9	0.61 (0.001)	2.9
wikivitals	10008	629521	126	2.4	0.26 (0.04)	2.7

they transfused in popular culture as the *six degree of separation* phenomenon. In fact, this phenomenon has since been empirically observed in many networks (see Watts, 2000; Newman, 2001b and Table 1.2).

Edge transitivity

A popular saying tells us that *"a friend of my friend is my friend"*. Thus, one would expect the interaction in a network to be *transitive*. This means that if Alice interacts with Bob, and Bob interacts with Cecile, then Alice and Cecile have a high probability of also being in interaction. The *clustering coefficient* measures this phenomenon. We define a connected triple as a set of three nodes, where one node is connected to two other nodes. We also define a triangle as a set of three nodes that are connected to each other. Since each triangle of three nodes contributes three connected triples (one centred on each of the three nodes), the clustering coefficient cc is given by

$$\text{cc} = \frac{3 \times \text{number of triangles}}{\text{number of connected triples of nodes}}.$$

Consider a graph in which the interactions among nodes are purely random (i.e., an interaction between two nodes occurs with a probability p). Since there are $\binom{n}{3}$ node sets of size three, the expected number of triangles is thus $\binom{n}{3}p^3$, and the expected number of connected triples is $\binom{n}{3}p^2$. Hence, the clustering coefficient of a random graph equals $3p$. Finally, since there are $\binom{n}{2}$ node pairs each which interact with probability p, then p can be estimated by the fraction $|E|/\binom{n}{2}$. Hence, the clustering coefficient of a random graph can be estimated by $\frac{6|E|}{n(n-1)}$. We observe in Table 1.2 that the clustering coefficients of real-world social networks are several orders of magnitude higher than those of random graphs of same size.

Heavy-tailed degree distribution

Let us denote by p_k the probability for a uniformly sampled node to have degree k and call $\{p_k : k = 0, 1, 2, \dots\}$ the *degree distribution*. In a random network, where $|E|$ edges are drawn uniformly at random among the $\binom{n}{2}$ node pairs, the degree distribution is binomial with parameters n, p, with $\hat{p} = |E|/\binom{n}{2}$ being an estimate for the edge probability. Nonetheless, in most networks the degree distribution is highly right-skewed, in other words, has a heavy tail distribution for values that are far above the mean. This highlights the fact that there are a small number of nodes having very large degrees (for example influencers in a social network), whereas the majority of nodes have very small degrees. Therefore, it is in general more accurate to model the degree distribution of real networks by a *power law*.

A random variable $X \in [x_{\min}, +\infty)$ is distributed according to a continuous power law of exponent α, if it is drawn from a probability distribution whose density is $f(x) = Cx^{-\alpha}$. While $\alpha > 1$ is required for the probability distribution to be well-defined (and then $C = (\alpha - 1)x_{\min}^{\alpha-1}$ from normalisation), typical values of α often lie in the range $2 < \alpha < 3$. An important property of power laws is that they are *scale-free* (or scale-invariant), namely $f(cx) \propto f(x)$ for any constant c. As the degrees are integer values, we will consider the discrete variant of a power law, namely the Zipfian distribution, where $\mathbb{P}(X = k) = Ck^{-\alpha}1(k \geq x_{\min})$ with $C = \left(\sum_{k=0}^{\infty} (k + x_{\min})^{-\alpha}\right)^{-1}$.

While fitting power laws is complex as large fluctuations occur in the tail of the distribution (Newman, 2005b; Clauset *et al.*, 2009), it is convenient to notice that $\log \mathbb{P}(X = k) = -\alpha \log k + \log c$ for $k \geq x_{\min}$, and thus with a log-log scale the probability distribution is a straight line. To reduce the effect of the aforementioned tail fluctuations, it is better to use the Complementary Cumulative Distribution Function (CCDF) for fitting instead of the density function. The Hill estimator also accurately estimates the exponent of a power law, see e.g., (Clauset *et al.*, 2009) for details. Figure 1.8 shows the power law of the *Citeseer* network.

While the power-law paradigm has been widely accepted and is sometimes referred to as a 'universal law', it has also been heavily criticized. In particular, a linear regression on the log-log plot generates significant systematic errors under relatively common conditions (see Clauset *et al.*, 2009, Appendix A). Moreover, Lima-Mendez and van Helden, 2009 showed that for biological networks the power-law degree distribution is a myth. Similarly, by applying goodness-of-fit tests on more than 1000 networks, Broido and Clauset, 2019 showed that networks with power-law degree distributions are actually rare. Nonetheless, a vast majority of real-world networks have *heavy-tailed* degree distributions.

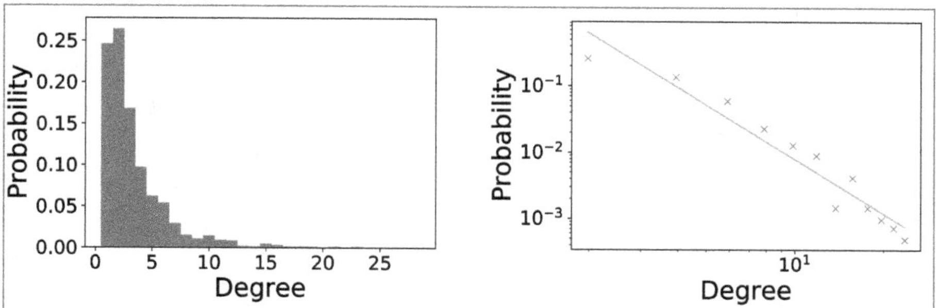

(a) Histogram of node degree. (b) Log-log plot with linear regression fitting.

Figure 1.8. The degree distribution of the *Citeseer* network.

1.2.2 How do these Properties Arise?

In order to explain how the described properties arise in networks, we introduce some *random graph models* with stochastic node interactions. The random graph models will be studied in detail in Chapter 2. These models will also serve as reference for studying statistical problems related to networks.

Erdős-Rényi random graphs

The simplest random graph model is the *Erdős-Rényi model*. This model has n nodes and each pair of nodes is connected with probability p.

This is a simple model, in particular since it assumes that interactions between different node pairs are independent. Hence, the model will not allow any of the edge transitivity. Morever, the degree distribution of an Erdős-Rényi random graph is binomial, $\text{Bin}(n, p)$,[5] which is not heavy tailed.

Nevertheless, the Erdős-Rényi model allows us to illustrate the properties of connectivity and sparsity in a beautiful manner. Indeed, since the degree distribution is binomial, it follows that the average degree $\bar{d}$ of the nodes is equal to np. If p is constant, then it means that $\bar{d}$ scales with the number of nodes n, and hence the graph is not sparse in this scaling regime. It is thus common to scale p with n, such that $p = p_n \ll 1$. For example, by choosing $p = \frac{a}{n}$ with a constant, we have $\bar{d} = a$, and the average degree remains constant as n grows. We will see in Chapter 2 that another interesting choice is $p_n = a\frac{\log n}{n}$, so that the average degree $\bar{d} = a \log n$ grows logarithmically with n. In Figure 1.9, two examples of Erdős-Rényi graphs are shown. We observe that when $p_n = \frac{2}{n}$ in (a), the graph is disconnected, *i.e.*, a significant number of nodes are grouped into one connected component, while

5. This is because a given node i has n potential neighbours ($n - 1$, if we exclude self-loops), and this node i is connected to another node with probability p.

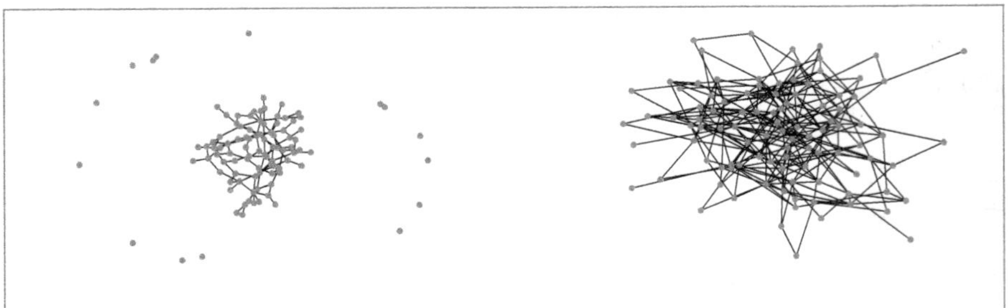

(a) $p_n = \frac{2}{n}$. (b) $p_n = \frac{2 \log n}{n}$.

Figure 1.9. Erdős-Rényi graphs with $n = 100$ and various interaction probabilities p_n.

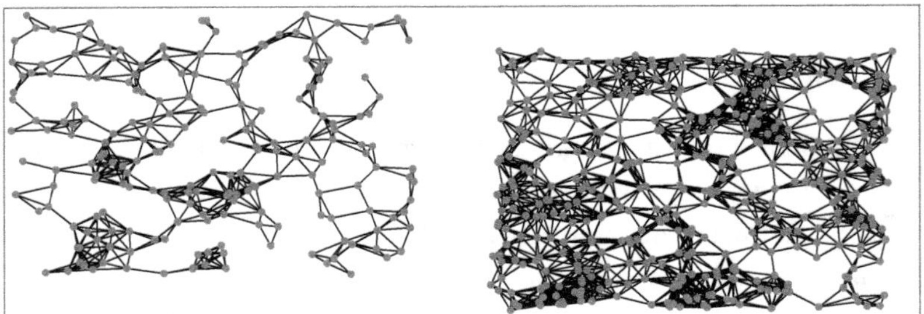

(a) $n = 200$ (b) $n = 400$

Figure 1.10. Example of RGG, when $\mathcal{S} = [0, 1]^2$ and $r = 0.1$, for different n.

some nodes remain isolated. On the contrary, when $p_n = \frac{2 \log n}{n}$ in (b), the graph appears to be connected. We will see in Chapter 2 how rigourous statements confirm these observations.

Random geometric graphs

Edge transitivity can be modelled by introducing *geometry*. Let us consider n nodes, and assume that each node has a random position on the Euclidean plane. Intuitively, nodes that are close to each other have more chance of being connected than nodes placed further apart. An extreme choice is to assume that two nodes are connected if and only if their Euclidean distance is less than a threshold r. This gives the *Random Geometric Graph* model. We observe in Figure 1.10 that this model leads to graphs with a large number of triangles (compared with Erdős-Rényi graphs). Moreover, the graphs appear locally dense while remaining globally fairly sparse.

Preferential attachment models

While the Erdős-Rényi model explains sparsity and connectivity, and geometric graphs explain transitivity, none of these models exhibit a power law degree distribution. To model networks with scale-free degree distributions, Solla Price, 1965,

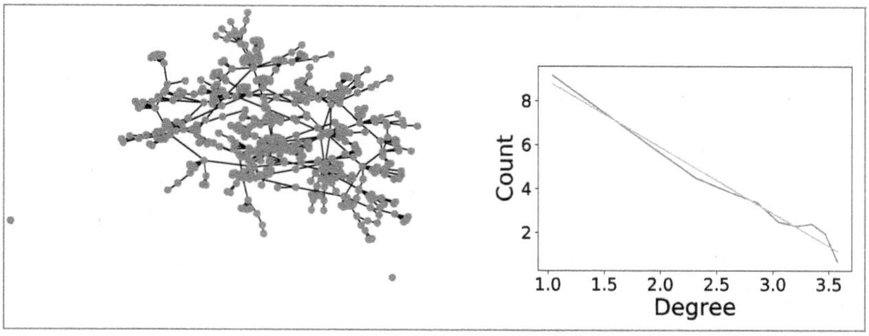

Figure 1.11. Left: A realisation of the preferential attachment model with $n = 500$ nodes. Right: Degree distribution in log-log scale of a preferential attachment graph with $n = 10^4$ nodes. The orange curve represents the linear regression fitting.

1976 (analysing citation networks) and Barabási and Albert, 1999 (analysing web-graphs) proposed the *preferential attachment model*. It is a growing network model in which a new node enters the network at each time step. The probability that the new node interacts with an existing node i is proportional to the degree d_i of node i. Therefore, nodes with large degree tend to attract new edges, hence increasing even more their degree. We plot an example of a graph generated by the preferential attachment model in Figure 1.11, as well as its degree distribution. We will give a rigorous definition of this model in Chapter 2 and prove that this model indeed has a power-law degree distribution in the limit.

1.3 What Are the Statistical Problems Related to Networks?

1.3.1 How to Cluster Network Nodes?

Community detection (also referred to as *community recovery* or *graph clustering*) is a very common problem in network analysis. It consists of grouping the nodes into K communities (also called groups, blocks or clusters), such that nodes inside a community have some similar properties. Intuitively, we shall assume that nodes in the same community are more likely to interact than nodes belonging to different communities.[6]

Again intuitively, a good partition should minimise the number of interactions between different clusters. Consequently, a first class of graph clustering methods, called *cut-based methods*, aim to find K clusters such that the number of interactions

6. This is sometimes referred to as *associative* communities. Nonetheless, some networks may be *dissasociative*. That is, interactions are more likely to occur between nodes in different communities.

between different clusters is minimised. It leads to several *spectral methods* that use the information contained in the eigenvectors of a well-chosen matrix to retrieve the communities. This beautifully links graph theory with linear algebra.

Other clustering methods assess the quality of a given partition via certain criteria which they aim to optimise. One example of this class of methods is based on the concept of *modularity*. In essence, modularity compares a graph with clusters to some reference random graph model. The maximisation of modularity is usually done via a greedy algorithm. One strength of such methods is that it is not necessary to know the number of clusters in advance.

Unfortunately, we will see that the modularity-based methods are prone to overfitting. In particular, we will show that on random graphs with no community structure, such as Erdős-Rényi random graphs, it is possible to find partitions with a high modularity! We will see how we can mitigate this problem by using *Bayesian* methods. Those methods assume that the graph data is generated from a random graph model with a clustering structure and look for the best parameters via a Markov Chain Monte Carlo algorithm.

1.3.2 Which Nodes are Most Important in a Network?

In large networks, many applications require the ranking of nodes in terms of importance. Examples include the identification of the most influential nodes in social networks, the study of super-spreaders of a disease, and the analysis of bottlenecks in urban or technological networks (such as electric grids). While all these problems are related to finding the most important, crucial nodes, the notion of importance varies greatly. Indeed, the most influential nodes in a social network may simply be the nodes with the largest degree. For example, when creating an account on *Twitter* or *Instagram*, the online social networks suggest the new users to follow popular users. On the contrary, bottlenecks in an electric grid are located on nodes with small degree such that, if those nodes were not in the network, there would be a great change in the network flow. Finally, other applications, such as PageRank, rank the nodes based on a random walk on the network nodes, mimicking browsing or searching behaviour.

1.3.3 How to Infer Important Information in a Network?

Analysing a very large network is often easier done via summary statistics. Some examples are: estimating the average age of users in a social network, finding the proportion of drug users in a population, polling before an election, etc. A first possibility is to uniformly sample k nodes, and average over this sample. Unfortunately, in practice, it is often hard to sample the nodes *uniformly*. Typically, uniform sampling in a huge social network like *Facebook* or *Twitter* cannot be done efficiently as (a) the list of all accounts on these platforms is not publicly available; and (b) there is a strict limitation on the API access rate. For instance, a standard *Twitter* account can make no more than one request per minute. At that rate, we would need about 950 years to crawl the entire Twitter social network...

Moreover, a small bias in the sampling process may lead to a very large bias in the estimator, as many famous examples involving polling before elections can attest. It is important to note that a bias in the node sampling cannot be mitigated by simply sampling more nodes. One infamous example involves *The Literary Digest*, who in 1936 had polled more than two million individuals and wrongly predicted a clear victory of Landon over Roosevelt. The way of sampling created a bias, since the newspaper simply polled over its own readers, who were wealthier than the average citizen.[7]

Book Organisation

The book is organised as follows. We start by presenting various random graph models in Chapter 2. Chapter 3 focuses on centrality indices in networks. Community detection problem is presented and analyzed in Chapter 4, and Chapter 5 is devoted to semi-supervised learning on networks, when some information about the community structure is given. In Chapter 6, we extend the community detection problem to temporal networks. Finally, in Chapter 7 we present techniques for sampling and performing questionaries in networks.

Book Bibliographic Position

Let us discuss the position of the book with respect to the other reference works. Random graph models are thoroughly analysed in Bollobás, 2001; Chung and Lu, 2006; Janson *et al.*, 2011; Hofstad, 2016. Graph formation processes (e.g., preferential attachment processes) and dynamics on graphs (e.g., epidemic processes) are studied in Durrett, 2007; Draief and Massoulié, 2010; Barabási, 2016; Newman, 2018; Masuda and Lambiotte, 2021. Specific applications of random graph and complex network models to social networks are discussed in Wasserman and Faust, 1994; Doreian *et al.*, 2005; Carrington *et al.*, 2005; Scott and Carrington, 2011; Prell, 2012; Yang *et al.*, 2016; Borgatti *et al.*, 2018; Knoke and Yang, 2019.

7. See https://en.wikipedia.org/wiki/The_Literary_Digest.

The fitting and visualization of random graphs and complex networks are presented in Ellson *et al.*, 2004; Hagberg *et al.*, 2008; Bastian *et al.*, 2009; Kolaczyk *et al.*, 2009; Goldenberg *et al.*, 2010; Cherven, 2015; Mrvar and Batagelj, 2016; De Nooy *et al.*, 2018; Kolaczyk and Csárdi, 2020. We do not cover the above topics in detail.

Our emphasis is on *fundamental statistical aspects* of complex network analysis (aka network science). Graph clustering and community detection, in particular clustering of stochastic block models, are studied in Newman, 2018; Abbe, 2018. This is still a very rapidly developing research area, with many interesting new results continuing to appear. Here, we summarise the main results in the community detection problem, review important progress since 2018 and study clustering in temporal networks. Semi-supervised learning is presented in Chapelle *et al.*, 2006. In this book, we focus on graph-based semi-supervised learning methods and their application to temporal networks.

To the best of our knowledge, there are no textbooks about the detailed analysis of network centrality indices (especially about their comparative analysis and their various applications beyond the scope of social networks). As is the case for the community detection problem, new important results continue to emerge. We have tried to do a state-of-the-art survey in this area. Also, we have not seen any textbook about modern methods for sampling in networks.

Thus, we hope that this is the first comprehensive textbook-style exposition of the statistical analysis of networks.

Funding

The work on this book was partly supported by Inria – Nokia Bell Labs Project "Distributed Learning and Control for Network Analysis" and EU COST Action "European Cooperation for Statistics of Network Data Science".

DOI: 10.1561/9781638280514.ch2

Chapter 2

Random Graph Models

This chapter is devoted to basic models for complex networks. We introduce several important classes of random graph models and we illustrate and study some statistical properties of these models, such as degree distribution and connectivity.

Notations In the following, $G = (V, E)$ denotes a graph, where $V = \{1, \ldots, n\}$ is the set of vertices (nodes) and E is the set of edges (links). We say that a graph G is a *random graph* if G was generated from a random graph model. A *random graph model* refers to a probability distribution over the set of all graphs.

We will denote by d_i the degree of node i. The vector $d = (d_1, \ldots, d_n)$ is called the *degree sequence* of the nodes. Given a random graph model, the degree d_i of a node i is a random variable and is distributed according to some probability distribution. When all the degrees are identically distributed (*i.e.*, $d_1, \cdots, d_n$ are all distributed according to the same probability distribution $\mathcal{D}$), we say that the degrees in the graph G are distributed according to the degree distribution $\mathcal{D}$.

2.1 Erdős-Rényi Random Graphs

2.1.1 Definition

Definition 2.1. Let n be an integer, and $P = (p_{ij})_{1 \leq i < j \leq n} \in [0;1]^{n \times n}$ be a set of probabilities. A *Bernoulli random graph* $G = (V, E)$ is an undirected, unweighted graph G such that:

- $V = \{1, \dots, n\}$;
- $\mathbb{P}\big((ij) \in E\big) = p_{ij}$ for all node-pair (i,j) with $1 \leq i < j \leq n$.

We write $G \sim \mathcal{G}(n, (p_{ij}))$. In a Bernoulli random graph, every node pairs (i,j) is connected by an edge with probability p_{ij}, independently of all other node pairs.

Remark 2.1. If $G \sim \mathcal{G}(n, (p_{ij})_{1 \leq i < j \leq n})$, then the adjacency matrix A of G is a symmetric random matrix, whose entries are independently distributed, with $A_{ij} = A_{ji} \sim \mathrm{Ber}(p_{ij})$ and $A_{ii} = 0$.

Proposition 2.1. *Let $G \sim \mathcal{G}(n, (p_{ij}))$ and A be its associated adjacency matrix. We have:*

$$\mathbb{P}(A) = \prod_{i<j} p_{ij}^{A_{ij}} (1 - p_{ij})^{1 - A_{ij}}.$$

Proof. The independence of the edge sampling process ensures that

$$\mathbb{P}(A) = \prod_{1 \leq i < j \leq n} \mathbb{P}\left(A_{ij}\right).$$

Moreover,

$$\mathbb{P}\left(A_{ij}\right) = \begin{cases} p_{ij} & \text{if } A_{ij} = 1 \\ 1 - p_{ij} & \text{if } A_{ij} = 0 \end{cases},$$

and this can conveniently be rewritten as $\mathbb{P}(A_{ij}) = p_{ij}^{A_{ij}} (1 - p_{ij})^{1 - A_{ij}}$. □

Example 2.1. Suppose that $\forall i, j : p_{ij} = p$. Then, $\mathcal{G}(n, (p_{ij}))$ is called the *Erdős-Rényi model*[1], and traditionally denoted by $\mathcal{G}(n, p)$ or $\mathcal{G}_{n,p}$.

1. This model was first introduced by Gilbert in 1959 (Gilbert, 1959), while the same year a paper from Erdős and Rényi study a similar but different model (Erdős and Rényi, 1959), where all graphs on a fixed vertex set with a fixed number of edges are equally likely. Asymptotically, these two models are equivalent.

Corollary 2.1. *Let $G \sim \mathcal{G}_{n,p}$ and A be its associated adjacency matrix. We have*

$$\mathbb{P}(A) = (1-p)^{\frac{n(n-1)}{2}} \left(\frac{p}{1-p}\right)^{|E|},$$

where $|E|$ is the number of edges of G.

Proof. Using Proposition 2.1, we can write

$$\mathbb{P}(A) = \prod_{i<j} p^{A_{ij}}(1-p)^{1-A_{ij}} = \prod_{i<j}(1-p)\left(\frac{p}{1-p}\right)^{A_{ij}}.$$

The result follows by noticing that $|E| = \sum_{i<j} A_{ij}$. □

Algorithm 1 provides a simple way to generate an Erdős-Rényi random graph, by looping over all possible node pairs (i,j), and adding (i,j) to the edge list with probability p.

Algorithm 1: Simple generation of Erdős-Rényi graphs.

```
Input: number of nodes n, edge probability p ∈ [0, 1].
Output: list of edges E.
Process:
E ← ∅;
for i = 1 to n-1 do
    for j = i+1 to n do
        x ← random number between 0 and 1;
        if x < p then
            add the edge (i, j) to E.
Return: E.
```

The space-complexity of Algorithm 1 is $O(|E|)$ (corresponds to storing $|E|$ edges), while its time-complexity is $O\left(n^2\right)$. In particular, it is very inefficient if p is small: indeed, in that case, the majority of node pairs (i,j) will not be connected, and we are wasting time by testing them. In other words, starting from node i, the node pairs $(i, i+1), \ldots, (i, i+k-1)$ will not be linked, while the pair $(i, i+k)$ will give an edge. This number k represents the number of failures in a sequence of independent Bernoulli random variables before the first success occurs. Hence, it is geometrically distributed with parameter p. Based on this observation, Batagelj and Brandes, 2005 proposed Algorithm 2 for an efficient generation of a sparse Erdős-Rényi graph. It has both space and time complexity of $O(|E|)$.

Algorithm 2: Fast generation of sparse Erdős-Rényi graphs.

```
Input: number of nodes n, edge probability p ∈ [0, 1].
Output: list of edges E.
Process:
E ← ∅;
i ← 0;
for i = 1 to n − 1 do
    v ← i
    while v ≤ n do
        k ← realisation of a geometric r.v. with parameter p;
        v ← v + k;
        if v ≤ n then
            add the edge (i, j) to E.
Return: E.
```

2.1.2 Degree Distribution

Proposition 2.2. *Let $G \sim \mathcal{G}(n,p)$, and let d_i be the degree of node i. Then d_i is distributed according to* $\mathrm{Bin}(n,p)$*. In particular, the average degree $\bar{d}$ of the graph equals np.*

Proof. Indeed, the degree of i, denoted d_i, is equal to $\sum_{j=1}^{n} A_{ij}$, where A_{ij} are i.i.d. Bernoulli random variable with parameter p. □

Remark 2.2. It has been observed that many real graphs have a heavy-tailed degree distribution (such as a power law), and not a binomial one (we refer to the discussion in Section 1.2). An intuitive argument is the following one: since binomial distributions are well concentrated, an Erdős-Rényi graph does not allow for many hubs (nodes with degrees much higher than the average degree), which we tend to see in real networks (*e.g.,* in a social network, some people will have many more connections than others and will act as influencers or hubs). Thus, the basic Erdős-Rényi random graph is not an appropriate model for many real networks.

2.1.3 Phase Transition Phenomena

Heuristic

This section considers sequences of Erdős-Rényi graphs $(G_1, \ldots, G_n, \ldots)$, such that G_n has n nodes, and the link-probability p_n depends on n. In other words,

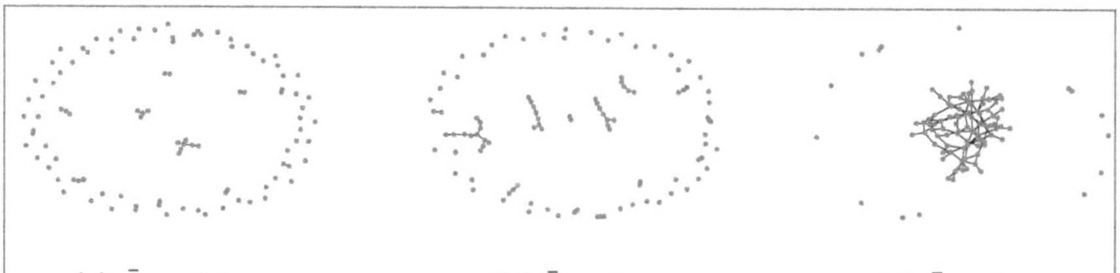

Figure 2.1. Erdős-Rényi graphs with $n = 100$ in the constant degree regime.

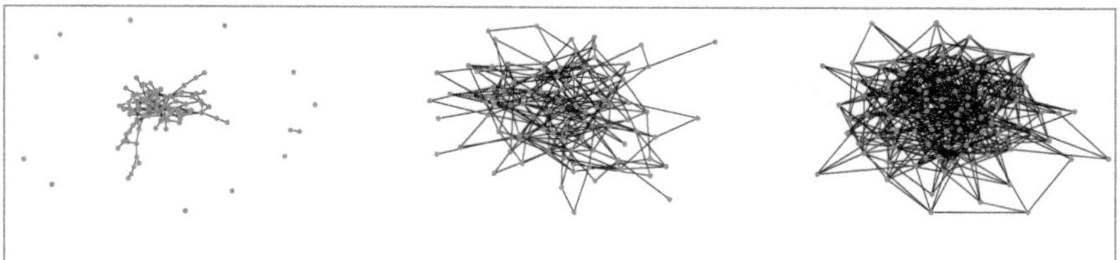

Figure 2.2. Erdős-Rényi graphs with $n = 100$ in the logarithmic degree regime.

$G_n \sim \mathcal{G}(n, p_n)$. We especially highlight two regimes:

- the regime $p_n = \frac{a}{n}$, where a is constant;
- the regime $p_n = a\frac{\log n}{n}$, where a is constant.

These two regimes are respectively called the constant degree regime and the logarithmic degree regime, as the expected degree $\bar{d}_n = np_n$ equals a in the first case and $a \log n$ in the second case.[2] Figures 2.1 and 2.2 show examples of Erdős-Rényi graphs in the constant and logarithmic degree regime respectively, for a given $n = 100$. We make the following observations in the constant degree regime:

- when $\bar{d}_n < 1$, most of the nodes are isolated, as expected since $\bar{d}_n < 1$ means that on average, a node has less than one neighbor;
- when $\bar{d}_n > 1$, it seems that there is a connected component which contains most of the nodes. We call this component the *giant component*.

On the other hand, in the logarithmic degree regime, we see that:

- if $\bar{d}_n < \log n$, the graph appears to be not connected, as there remains some isolated nodes or isolated edges;
- on the contrary, when $\bar{d}_n > \log n$, the graph appears to be fully connected.

Those observations are further strengthened by Figure 2.3. In the constant degree regime $p_n = \frac{a}{n}$, Figure 2.3(a) shows that when $a < 1$, the proportion of nodes in

2. More exactly, $\bar{d}_n = (n-1)p_n$ but if we allow self-loops, we can write $\bar{d}_n = np_n$, and moreover the difference is negligible for large n.

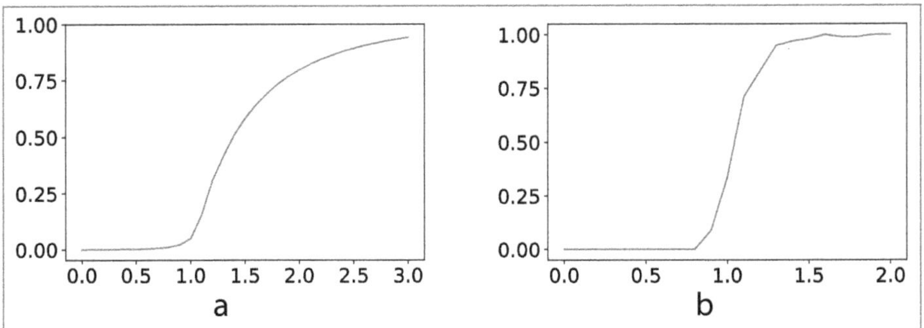

(a) Fraction of the nodes belonging to the largest connected component when $d_n = a$.

(b) Empirical probability that the graph is connected when $d_n = a \log n$.

Figure 2.3. Empirical evidence of a phase transition for the existence of a giant component in the constant degree regime and a phase transition for connectivity in the logarithmic degree regime of Erdős-Rényi graphs (here $n = 5000$).

the largest connected component is tiny. But, as soon as $a > 1$, this proportion becomes non-negligible, and increases steadily with a. Similarly, in the logarithmic degree regime $p_n = a\frac{\log n}{n}$, Figure 2.3(b) shows that the empirical probability that the graph is connected goes from 0 to 1 as soon as a becomes larger that 1.

Main statements

Let us now present two main statements justifying our previous heuristic observations.

Theorem 2.1 (Phase transition for giant component – constant degree regime). *Let $G \sim \mathcal{G}(n, p_n)$ be an Erdős-Rényi graph, with $p_n = \frac{a}{n}$ where a is a constant. Almost surely, the following holds:*

(a) *if $a < 1$, then there is no connected component of size larger than $O(\log n)$;*
(b) *if $a = 1$, then there is one large component of size $O(n^{2/3})$;*
(c) *if $a > 1$, then there is one and only one component of size $O(n)$. This component is called the* giant component.

The proof of Theorem 2.1 is complex and will not be presented in this book. We refer the interested reader to (Hofstad, 2016).

Theorem 2.2 (Phase transition for connectivity). *Let $G_n \sim \mathcal{G}(n, p_n)$ be a sequence of Erdős-Rényi random graphs. Let $\bar{d}_n = np_n$. The following holds.*

(a) *If there exists a sequence $(\omega_n)_n$ with $\omega_n \to +\infty$ such that $d_n < \log n - \omega_n$, then G_n is a.s. non connected. More precisely, the graph G_n contains a.s. at least one isolated node;*
(b) *If there exists a sequence $(\omega_n)_n$ with $\omega_n \to +\infty$ such that $\bar{d}_n > \log n + \omega_n$, then G_n is a.s. connected.*

Example 2.2. Assume $\bar{d}_n = \log n + \log\log n$, and let $G_n \sim \mathcal{G}(n, p_n)$. Then, Theorem 2.2 states that asymptotically G_n will be a.s. connected (we can take $\omega_n = \log\log n$).

Example 2.3. If $\bar{d}_n = a \log n$, with a constant, then Theorem 2.2 applies with $\omega_n = (a-1)\log n$. Hence G_n will be connected if $a > 1$, and will be disconnected if $a < 1$. In particular, this justifies the heuristic observation from Figure 2.3(b).

Proof of the connectivity phase transition

Before proving Theorem 2.2, let us start with the following lemma about the presence of isolated nodes in an Erdős-Rényi graph.

Lemma 2.4. *The probability that an Erdős-Rényi graph $G_n \sim \mathcal{G}(n, p_n)$ contains at least one isolated node satisfies*

$$\lim_{n\to\infty} \mathbb{P}\big(\exists \textit{ isolated node}\big) = \begin{cases} 0 & \text{if } p_n \geq \frac{\log n + \omega_n}{n} \text{ for some } \omega_n \to +\infty, \\ 1 & \text{if } p_n \leq \frac{\log n - \omega_n}{n} \text{ for some } \omega_n \to +\infty. \end{cases}$$

This lemma implies that if $p_n \leq \frac{\log n - \omega_n}{n}$, then the graph contains a.s. an isolated node, and hence the graph is a.s. not connected. This precisely corresponds to the statement (a) of Theorem 2.2.

Proof of Lemma 2.4. Denote by A_i the event that "node i is isolated", and let $I_n = \sum_{i=0}^{n} 1(A_i)$ be the number of isolated nodes. Recall that $\bar{d}_n = np_n$ is the mean degree. We have

$$\mathbb{P}(A_i) = (1-p_n)^{n-1} = \left(1 - \frac{\bar{d}_n}{n}\right)^{n-1} \sim \exp\left(-\bar{d}_n\right) \sim \frac{1}{n}\exp(\mp\omega_n),$$

and thus

$$\mathbb{E}\big(I_n\big) = \sum_{i=0}^{n} \mathbb{P}(A_i) = n\mathbb{P}(A_1) \sim \mathrm{e}^{\mp\omega_n}.$$

(i) If $\bar{d}_n = \log n + \omega_n$, we have $\mathbb{E}\big(I_n\big) \sim \mathrm{e}^{-\omega_n} \to 0$. Since the expected number of isolated nodes goes to 0, we can conclude the proof using the first moment method. Indeed, recall that by Markov inequality (see Proposition A.6 and Corollary A.2), we have:

$$\mathbb{P}\big(\exists \text{ isolated node}\big) = \mathbb{P}(I_n \geq 1) \leq \frac{\mathbb{E}I_n}{1} \longrightarrow 0.$$

(ii) If $\bar{d}_n = \log n - \omega_n$, we have $\mathbb{E}\big(I_n\big) \sim \mathrm{e}^{+\omega_n} \to +\infty$, and hence the expected number of isolated nodes goes to infinity. Unfortunately, this is not enough to conclude anything about the probability of existence of an isolated node, and we will need the second moment method. Indeed, we have to show that the random variable I_n is well-concentrated around its mean. Since its mean diverges to infinity, the result will follow. For that, we will use Chebyshev's inequality (Proposition A.7). We have

$$\mathrm{Var}\,(I_n) = \mathbb{E}\left(I_n^2\right) - (\mathbb{E}I_n)^2 .$$

Note that

$$\begin{aligned}\mathbb{E}\left(I_n^2\right) &= \mathbb{E}\left(\sum_i \sum_j 1(A_i)1(A_j)\right)\\ &= \sum_i \sum_j \mathbb{P}\left(A_i, A_j\right)\\ &= n\mathbb{P}\big(A_1\big) + n(n-1)\mathbb{P}\big(A_1 \cap A_2\big).\end{aligned}$$

Here we need to be careful, since A_1 and A_2 are not independent. Indeed, knowing that node 1 is isolated means that there is no edge between node 1 and node 2, and thus weakly increases the probability that node 2 is isolated. We have:

$$\begin{aligned}\mathbb{P}\big(A_1 \cap A_2\big) &= \mathbb{P}\big(A_2 \,|\, A_1\big)\, \mathbb{P}\big(A_1\big)\\ &= (1-p_n)^{n-2}\, \mathbb{P}\big(A_1\big)\\ &= \frac{1}{1-p_n}\left(\mathbb{P}\big(A_1\big)\right)^2,\end{aligned}$$

since $\mathbb{P}\big(A_1\big) = (1-p_n)^{n-1}$. Lastly,

$$\begin{aligned}(\mathbb{E}I_n)^2 &= \left(\sum_i \mathbb{P}\left(A_i\right)\right)^2\\ &= \sum_i \sum_j \mathbb{P}(A_i)\mathbb{P}(A_j)\\ &= \sum_i \sum_j \mathbb{P}(A_1)^2\\ &= n^2\, \mathbb{P}(A_1)^2.\end{aligned}$$

Putting all pieces together leads to

$$\begin{aligned}\mathrm{Var}(I_n) &= n\mathbb{P}\big(A_1\big) + n(n-1)\mathbb{P}\big(A_1 \cap A_2\big) - n^2\mathbb{P}(A_1)^2\\ &= n\mathbb{P}\big(A_1\big) + \frac{n(n-1)}{1-p_n}\mathbb{P}\big(A_1\big)^2 - n^2\mathbb{P}\big(A_1\big)^2\\ &\leq n\mathbb{P}\big(A_1\big) + \frac{n^2}{1-p_n}\mathbb{P}\big(A_1\big)^2 - n^2\mathbb{P}\big(A_1\big)^2\end{aligned}$$

$$= n\mathbb{P}(A_1) + n^2\mathbb{P}(A_1)^2\Big(\frac{1}{1-p_n} - 1\Big)$$

$$= \mathbb{E}\,(I_n) + (\mathbb{E}I_n)^2\,\frac{p_n}{1-p_n}.$$

Thus, by the second moment method (see Corollary A.5)

$$\mathbb{P}(I_n = 0) \;\le\; \frac{\mathrm{Var}(I_n)}{\big(\mathbb{E}(I_n)\big)^2} \;\le\; \frac{1}{\mathbb{E}(I_n)} + \frac{p_n}{1-p_n}.$$

Since $\mathbb{E}I_n \to \infty$ and $p_n \to 0$, this last quantity goes to zero when n goes to infinity. □

We can now prove the part (b) of Theorem 2.2.

Proof of Theorem 2.2(b). Suppose that $d_n \ge \log n + \omega_n$. Then, Lemma 2.4 shows that the number of isolated nodes I_n is zero. To show that G_n is indeed connected, we need to show that

$$\mathbb{P}\,(G_n \text{ is disconnected and } I_n = 0) \to 0.$$

If G_n is disconnected and has no isolated nodes, then G_n contains a connected component C_k of size $2 \le k \le \lfloor n/2 \rfloor$. Directly counting the expected number of components of size k is difficult, as the probability of them depends on the exact number of edges they contain (which can be as low as $k-1$ if C_k is a tree, up to $\frac{k(k-1)}{2}$ if C_k is complete). To avoid this issue, we will notice that the component C_k contains *spanning trees*. By spanning tree of C_k, we mean sub-graph of C_k which is a connected tree that contains all the vertices of C_k. Note that C_k can contain more than one spanning tree.

Let us denote by X_k the number of spanning trees of size k. By the previous observation, X_k is larger than the number of connected components of size k. Moreover, if G_n is disconnected and has no isolated nodes, then there must be a $k \in \{2, \ldots, \lfloor n/2 \rfloor\}$ such that $X_k \ge 1$. Hence by the union bound and the first moment method,

$$\begin{aligned}
\mathbb{P}\,(G_n \text{ is disconnected and } I_n = 0) \;&\le\; \mathbb{P}\left(\bigcup_{k=2}^{\lfloor n/2 \rfloor} \{X_k \ge 1\}\right)\\
&\le\; \sum_{k=2}^{\lfloor n/2 \rfloor} \mathbb{P}\,(X_k \ge 1)\\
&\le\; \sum_{k=2}^{\lfloor n/2 \rfloor} \mathbb{E}X_k. \qquad (2.1)
\end{aligned}$$

We need to bound $\mathbb{E}X_k$. Firstly, there is $\binom{n}{k}$ ways of choosing k vertices $(v_1, \ldots, v_k)$ among the n nodes. Once these k vertices chosen, then by Cayley's theorem [see Theorem 3.17 of Hofstad, 2016], there is possibly k^{k-2} trees containing those vertices. Since those k vertices form a tree within G_n, they are linked by $k-1$ edges, which has a probability p_n^{k-1} of occurring. Lastly, the graph G_n should not include any edge between the tree and the rest of the graph: this has a probability $(1-p_n)^{k(n-k)}$. To summarize,

$$\mathbb{E}X_k = \binom{n}{k} k^{k-2} p_n^{k-1} (1-p_n)^{k(n-k)}.$$

Applying the Stirling bound $k! \geq k^k \mathrm{e}^{-k}$, we have $\binom{n}{k} \leq (n\mathrm{e}/k)^k$. Moreover $(1-p_n)^{k(n-k)} \leq \mathrm{e}^{-p_n k(n-k)} \leq e^{-knp_n/2}$ and $np_n \geq 1$. Thus

$$\mathbb{E}X_k \leq n\frac{\mathrm{e}}{k^2}\left(np_n\mathrm{e}\right)^{k-1}\mathrm{e}^{-knp_n/2} \leq n\left(np_n\mathrm{e}^{1-np_n/2}\right)^k.$$

Note that the function $f(x) = x\mathrm{e}^{1-x/2}$ is decreasing for $x \geq 2$. Since $np_n = \log n + \omega_n$, we have for n large enough $np_n \geq \log n$. Hence

$$\mathbb{E}X_k \leq n\left(\log n\mathrm{e}^{1-\log n/2}\right)^k \leq n\left(\frac{\mathrm{e}\log n}{2\sqrt{n}}\right)^k,$$

and for any $m \geq 1$,

$$\sum_{k=m}^{\lfloor n/2 \rfloor} \mathbb{E}X_k \leq n\left(\frac{\mathrm{e}\log n}{2\sqrt{n}}\right)^m \left(\frac{1}{1-\frac{\mathrm{e}\log n}{2\sqrt{n}}}\right) \leq 2n\left(\frac{\mathrm{e}\log n}{2\sqrt{n}}\right)^m$$

using $\frac{e\log n}{2\sqrt{n}} \leq \frac{1}{2}$ for n large enough. The previous bounding is rough, but enough to show that $\sum_{k=2}^{\lfloor n/2 \rfloor} \mathbb{E}X_k$ converges to zero. Showing that $\mathbb{E}X_1$ goes to zero is immediate, and going back to Equation (2.1) it follows that

$$\mathbb{P}\left(G_n \text{ is disconnected and } I_n = 0\right) \to 0.$$

This proves the statement (b) of Theorem 2.2. □

2.2 Other Random Graph Models

2.2.1 Configuration Model

In this section, we aim to construct a random graph G_n fitting a given degree sequence $d = (d_1, \ldots, d_n)$. This means that the graph G_n should have n nodes, and the edges are drawn such that node i has degree d_i. Let us make a few remarks.

- We can suppose $d_i \geq 1$, as $d_i = 0$ means that node i is isolated.
- It is not obvious that there exists a graph verifying the degree requirement. In fact, such a graph does not necessarily exist. For example, if we assume that the graph is unweighted, then $\sum_{i=1}^{n} d_i$ should be even (since this sum corresponds to two times the number of edges).
- Even if we assume that $\sum_{i=1}^{n} d_i$ is even, the construction of such a graph might not always be possible. To avoid those issues, we will allow self-loops and multi-edges.

Definition 2.2 (Configuration model)**.** Let $d = (d_1, \ldots, d_n)$ be a sequence such that $\sum_{i=1}^{n} d_i$ is even. At each node $i \in \{1, \ldots, n\}$, we attach d_i *half-edges* (also called *stubs*). We then pair the stubs by pairs of two, uniformly at random. The resulting graph is called *configuration model* with degree sequence d, abbreviated as $\mathrm{CM}_n(d)$.

This model allows multi-edges and self-loops. Moreover, by convention a self-loop counts for two in the degree of a node, since it comes from two stubs. Algorithm 3 generates a configuration model graph.

Algorithm 3: Generation of a configuration model graph.

```
Input: degree sequence (d_1, ..., d_n).
Output: list of edges E.
Process:
if Σ_{i=1}^{n} d_i is odd then
    return an error.
else
    E ← ∅;
    L ← ∅;
    for i = 1 to n do
        for k = 1 to d_i do
            add i to the list L.
    shuffle the elements of L;
    j ← 0;
    while j ≤ |L| do
        add the edge (L[j], L[j + 1]) to E;
        j ← j + 2.
Return: E.
```

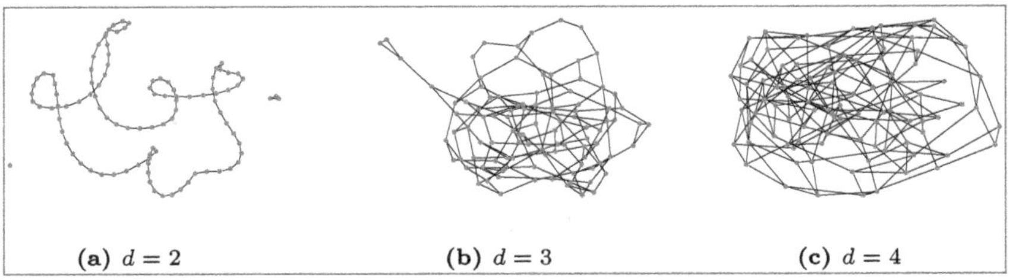

Figure 2.4. (n, d)-random regular graphs for $n = 100$ and various d.

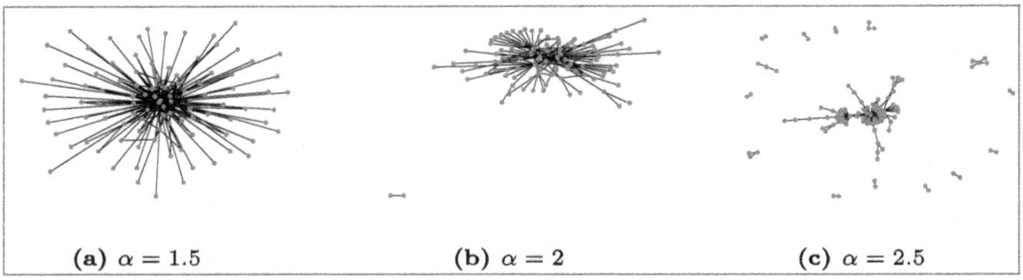

Figure 2.5. Configuration model with $n = 100$, where the degrees d_i are sampled independently from a Zipfian distribution with exponent α.

Example 2.4. If $d_1 = \cdots = d_n = d$, then we obtain a *random (n, d)-regular graph* (*i.e.*, a random graph with n nodes where all the nodes have the same degree d). We plot some examples in Figure 2.4.

Example 2.5. A random variable X follows a *Zipfian distribution* with parameters n and s if $X \in \{1, \cdots, n\}$ almost surely and $\mathbb{P}(X = k) = C^{-1}k^{-s}$ for $k \in [n]$, where $C = \sum_{k=1}^{n} k^{-s}$ is a normalisation constant. In Figure 2.5 we plot some graphs drawn from the configuration model, where the d_i are sampled from Zipfian distributions.

2.2.2 Preferential Attachment Model

Motivation

Previous models are static, in the sense that the number of nodes is fixed. Moreover, they do not explain how interesting properties (heavy-tailed degree distribution, etc.) arise in real graphs. This section provides an example of growing random graphs, where nodes and edges are added over time.

A first possibility is to construct a graph sequence $(G_n)_{n\in\mathbb{N}}$ such that each G_n is an Erdős-Rényi graph $\mathcal{G}(n, p)$. The graph G_{n+1} would be constructed from G_n as follow. The edges in G_n are copied to G_{n+1}, while edges of the form $(i, n + 1)$ (for $i = 1, \cdots, n$) are added independently with probability p. The new graph G_{n+1} is thus a $\mathcal{G}_{n+1,p}$, and G_n is a sub-graph of G_{n+1}. The problem is that the degree sequence is binomial, hence does not fit what we observe in most of real networks.

The *preferential attachment paradigm* offers an intuitive explanation behind the power law degree distribution that we seem to observe in reality. In this paradigm, a new node $n+1$ will be connected to the n existing nodes by some additional edges. Theses new edges $(i, n+1)$ are drawn independently with a probability proportional to the degree of the vertex i at that time. Thus, the new node $n+1$ is more likely to be connected to a node with a large degree.

Definition 2.3 (Preferential attachment – Informal definition)**.** At time t, a new node will be connected to an old node i with a probability proportional to the degree $d_i(t)$ of the old node (at time t).

With that definition, we can make the following remarks:

- the old nodes will tend to have higher degrees than the new ones;
- *the rich gets richer phenomenon*: new nodes tend to be attached to high degree old nodes. In particular, we expect the formation of hubs.

The fact that the graph will have hubs tend to make us think that the degree distribution will not be binomial, but may instead exhibit a power law. We will establish this in Proposition 2.3, just after giving a careful definition of the model.

Remark 2.3. The term *preferential attachment* comes from Barabási and Albert, 1999, who proposed a similar model, albeit not rigorously defined. Their model was actually close to the older works by Yule, 1925 and Solla Price, 1976. For a fully rigorous treatment, we refer to Bollobás *et al.*, 2001 and Hofstad, 2016.

Model definition

Definition 2.4. A sequence of graphs $\left\{G_t = (V_t, E_t), t \in \mathbb{N}\right\}$ is said to be drawn from the *Preferential Attachment Model* if:

- $|V_1| = 1$ and $|E_1| = 1$: at time step $t = 1$, we have one node v_1 with a single self-loop;
- at time step $t+1$, we add the node v_{t+1} to the graph. This node will be linked to one (and only one) node. The probability that the new node is connected to node v_i is given by

$$\mathbb{P}\Big((v_{t+1}, v_i) \in E_{t+1} \Big| G_t\Big) = \begin{cases} \frac{1}{2t+1} & \text{if } v_i = v_{t+1} \\ \frac{d_i(t)}{2t+1} & \text{otherwise,} \end{cases} \tag{2.2}$$

  where $d_i(t)$ is the degree of node v_i at time t (recall that by convention, a self-loop increases the degree by 2).

We present in Figure 2.6 some graphs drawn from the preferential attachment model.

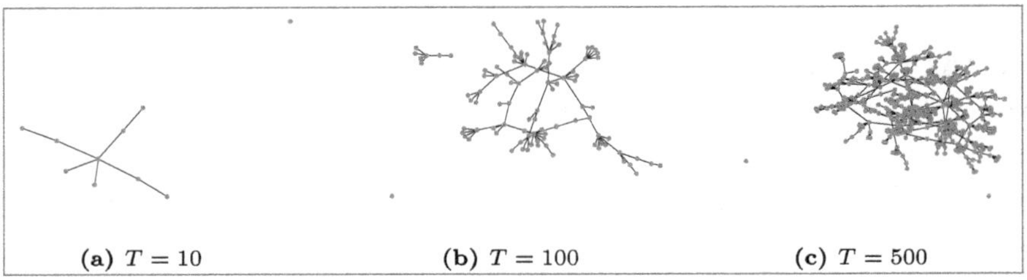

Figure 2.6. Graphs drawn from the preferential attachment model, for various T.

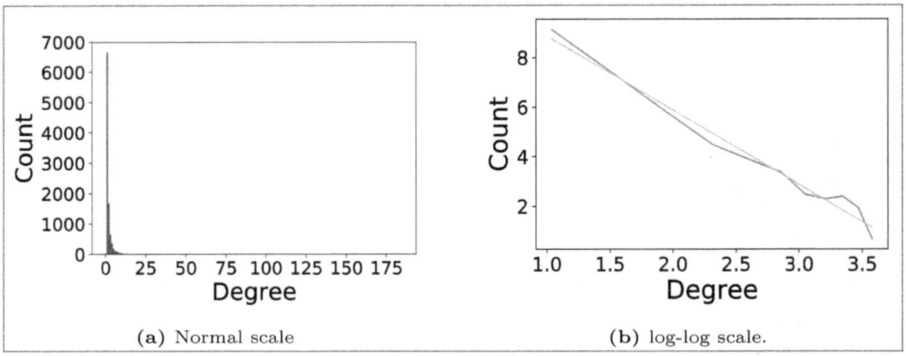

Figure 2.7. Degree distribution of a graph drawn from the preferential attachment model, with $T = 10^4$. Left: normal scale. Right: log-log scale. The orange slope represents the curve $y = -3x + 11.9$ obtained by linear regression fitting.

Lemma 2.5. *After t time-steps, the preferential attachment model results in a network with* $|V_t| = t$ *nodes and* $|E_t| = t$ *edges. In particular, Equation* (2.2) *defines a probability distribution.*

Proof. Indeed, at each time step, we add one node, so $|V_t| = t$. Moreover, we add only one edge per time step. Last, since $\sum_{i=1}^{t} d_i(t) = 2|E_t| = 2t$, then $\sum_{i=1}^{t+1} \mathbb{P}\Big((v_{t+1}, v_i) \in E_{t+1} \Big| G_t\Big) = 1$. □

Remark 2.4. A more general version of the preferential attachment model is described in Hofstad, 2016. Definition 2.4 corresponds to the case $m = 1$ and $\delta = 0$ there.

Degree distribution of the preferential attachment model

Let us now investigate the degree distribution of a graph drawn from the preferential attachment model. Figure 2.7 shows the histogram of the degrees. In particular, we see that in a log-log scale the curve seems to be linear. Let N_k be the number of nodes having degree k. Figure 2.7(b) seems to indicate that $\log N_k = -\alpha \log k + C$ where $\alpha = -3$ and C is a constant. This in turn implies $N_k \propto k^{-3}$, i.e., the degree distribution follows a power law with exponent 3. This is indeed proved in Proposition 2.3.

Proposition 2.3. *When $t \to +\infty$, the preferential attachment model exhibits a power law degree distribution with exponent* 3.

Proof. Let $s \in \{1, \dots, t\}$, and denote by $p(k, s, t)$ the probability that the vertex v_s has degree k at time t. The evolution of $p(k, s, t)$ is described by the *master equation*

$$p(k, s, t+1) = \frac{k-1}{2t+1} p(k-1, s, t) + \left(1 - \frac{k}{2t+1}\right) p(k, s, t), \tag{2.3}$$

with the initial condition $p(k, 1, 1) = \delta_{k,1}$ and the boundary condition $p(k, t, t) = \delta_{k,1}$. The term $\frac{k-1}{2t+1}$ represents the probability that the new node v_{t+1} is linked to node v_s at time $t+1$ (thus increasing the degree of s by 1), and $\left(1 - \frac{k}{2t+1}\right)$ is the probability that the new node v_{t+1} is not linked to node v_s.

Let $P(k, t)$ denote the total degree distribution of the entire network, that is the average of $p(k, s, t)$ over all nodes $v_s \in [t]$ present at time t. We have

$$P(k, t) = \frac{1}{t} \sum_{s=1}^{t} p(k, s, t).$$

Using equation (2.3), yields

$$(t+1)P(k, t+1) = \frac{k-1}{2t+1} tP(k-1, t) + \left(1 - \frac{k}{2t+1}\right) tP(k, t).$$

Therefore, the time evolution of $P(k, t)$ can be written as

$$(t+1)P(k, t+1) - tP(k, t) = \frac{t}{2t+1}\Big((k-1)P(k-1, t) - kP(k, t)\Big) + \delta_{k,1}.$$

When $t \to +\infty$, this equation for the stationary distribution reduces to

$$P(k) + \frac{1}{2}\Big(kP(k) - (k-1)P(k-1)\Big) = \delta_{k,1},$$

where $P(k)$ stands for $\lim_{t \to +\infty} P(k, t)$. This last equation is the discrete version of the differential equation

$$P(k) + \frac{1}{2} \frac{\mathrm{d}kP(k)}{\mathrm{d}k} = 0,$$

whose solution is

$$P(k) = Ck^{-3},$$

with a normalisation factor C such that $\sum_k P(k) = 1$ (*i.e.*, $C = \sum_{k=1}^{\infty} k^{-3}$). □

Remark 2.5. The above proof is not totally rigorous, as it involved some approximations that need rigorous justification. However, it explains well the essence of the preferential attachment process. We refer to Hofstad, 2016 for a more mathematically involved (but rigorous) proof, as well as some other deeper results on the preferential attachment model. In particular, more involved models (with the new nodes attached to several nodes or/and with several new nodes at each time step) result in power laws with various exponents.

2.2.3 Spatial Networks: Random Geometric Graphs, etc

In many situations, nodes are positioned in a metric space (*e.g.,* $\mathbb{R}^2$ or the sphere $\mathcal{S}_2$), and an interaction between two nodes directly depends on how far away the two nodes are in this space. Examples include base stations in wireless and sensors networks, in which two devices will be connected if they are not too far from each other. Moreover, in many networks, nodes possess attributes or features (*e.g.,* gender, age, a grade, a type, ...) which can also be represented as a position in a metric space, and influence link formation. For example in social networks, users of similar age and/or gender are typically more connected.

Definition 2.5. The *Spatially Embedded Random Network* (SERN) model is defined as follows. Let $(\mathcal{S}, d)$ be a metric space, and $(X_1, \ldots, X_n)$ be a random vector representing the locations of n nodes in $\mathcal{S}$. Let $\gamma : \mathbb{R}^+ \to [0, 1]$ be a *connectivity function*. Then, for every node pair (i, j), we draw an undirected edge between i and j with probability $\gamma\left(d(X_i, X_j)\right)$, where $d(X_i, X_j)$ denotes the distance between nodes i and j.

Example 2.6. In the *Random Geometric Graph* model (RGG) it is assumed that $X_1, \ldots, X_n$ are i.i.d. and uniformly distributed in $\mathcal{S}$, while $\gamma(x) = 1(x \leq r)$. In other words, two nodes are connected if and only if the distance between them is less than some threshold r.

We plot in Figure 2.8 some examples of RGG. We notice that when n is large, the network is composed of a few densely connected parts, with empty regions between them. Moreover, the graph is not small-world, as it takes a lot of edges to join two nodes that are far away from each other.

Example 2.7. The *Waxman model* is a SERN where X is uniformly distributed in $\mathcal{S}$, and $\gamma(x) = \min\left(1, qe^{-\alpha x}\right)$ where $q, \alpha > 0$ are some parameters.

Figure 2.9 shows some realisations of Waxman graphs, and we observe different behaviour than that of RGG. In particular, Waxman graphs look like small-world networks. Indeed, and in contrast to random geometric graphs, nodes that are far away can still be connected with a small but non-zero probability.

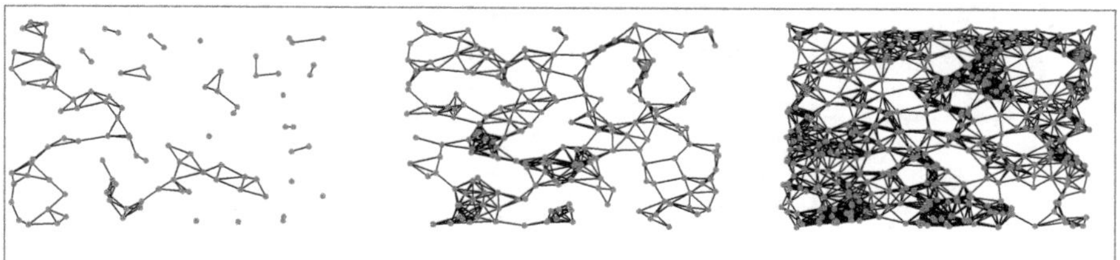

(a) $n = 100$ (b) $n = 200$ (c) $n = 400$

Figure 2.8. Example of RGG, when $\mathcal{S} = [0,1]^2$ and $r = 0.1$, for different n.

(a) $n = 100$ (b) $n = 200$ (c) $\gamma(x) = \min(1, 0.1e^{-5x})$

Figure 2.9. Examples of Waxman graphs, when $\mathcal{S} = [0,1]^2$, $q = 0.1$ and $\alpha = 5$, for different n. Figure (c) shows the connectivity function $\gamma(x) = \min(1, qe^{-\alpha x})$.

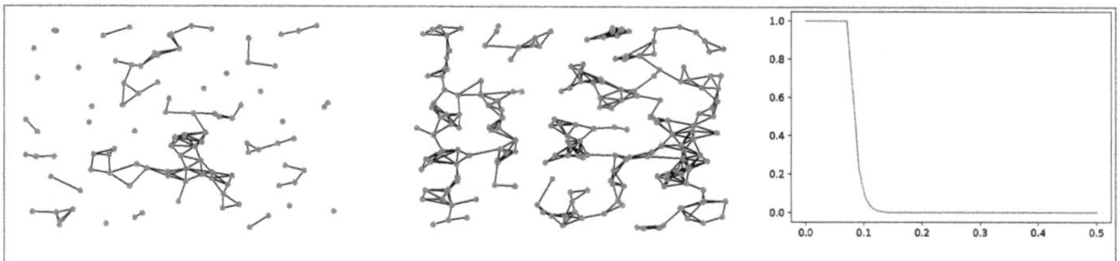

(a) $n = 100$ (b) $n = 200$ (c) $\gamma(x) = \min(1, qe^{-\alpha x})$

Figure 2.10. Examples of Waxman graphs, when $\mathcal{S} = [0,1]^2$, $\alpha = 100$ and $q = 2200$, for different n. q is chosen to be approximately equals to $e^{\alpha * 0.1}$, hence making the graphs look like RGG with cutoff $r = 0.1$. Figure (c) shows the connectivity function $\gamma(x) = qe^{-\alpha x}$, which indeed looks like $x \mapsto 1(x \leq r)$.

Finally, a RGG with threshold r can be expressed as a limit of a Waxman model, when $\alpha \to \infty$ and $q = e^{\alpha r}$ (see Figure 2.10).

In spatial networks as defined in Definition 2.5, while the random variables $(A_{ij})_{i<j}$ are still pair-wise independent, they are in general no more mutually independent. Indeed, the presence of an edge between i and j and between j and k influences the probability of an edge between i and k. The simplest example is to consider a random geometric graph. Knowing that $A_{ij} = A_{jk} = 1$, implies $d(X_i, X_j) \leq r$ and $d(X_j, X_k) \leq r$, and therefore the triangular inequality implies $d(X_i, X_k) \leq 2r$, i.e., node k cannot be arbitrarily far away from node i, increasing the likelihood of having an edge between i and k.

Table 2.1. Basic properties verified by the models presented in this chapter. Note that many of those properties only hold under specific conditions (see *e.g.*, Theorems 2.1 and 2.2), and this table is only for a rough summary purpose.

	Erdős-Rényi	CM	PA	RGG
Connectivity / Giant component	✓	✓	✓	✓
Small world	✓	✓	✓	×
Power law degree distribution	×	✓	✓	×
Edge transitivity	×	×	×	✓

2.2.4 Summary

We summarise in Table 2.1 the basic properties verified by the random graph models presented in this chapter.

2.3 Clustered Random Graphs: Block Models

This section is devoted to *clustered random graph models*. This refers to situations in which each node has a community attribute, and these community attributes influence the probability of interaction. The *block model* paradigm considers that nodes are placed into communities (called blocks) and that the probability of a link between i and j depends on the community labels of i and j (and eventually on some extra features of i and j, such as their spatial position).

2.3.1 Stochastic Block Model

The *Stochastic Block Model* is the simplest and most studied clustered random graph. It is a direct extension of the Erdős-Rényi model.

Definition 2.6. Let n be the number of nodes, K be the number of communities, $\pi = (\pi_1, \ldots, \pi_K)$ be a probability vector, and P be a $K \times K$ symmetric matrix whose entries are in $[0, 1]$. The pair (z, G) is drawn under the *Stochastic Block Model* (SBM) with parameters (n, π, P) if:

- $z \in [K]^n$ is a random vector whose entries are independent and identically distributed such that $\mathbb{P}(z_i = k) = \pi_k$;
- G is an undirected graph with n nodes, where the nodes i and j are connected with probability $P_{z_i z_j}$ independently of other pairs of nodes.

We write $(z, G) \sim \mathrm{SBM}(n, \pi, P)$.

Figure 2.11 gives some examples of graphs drawn from the SBM.

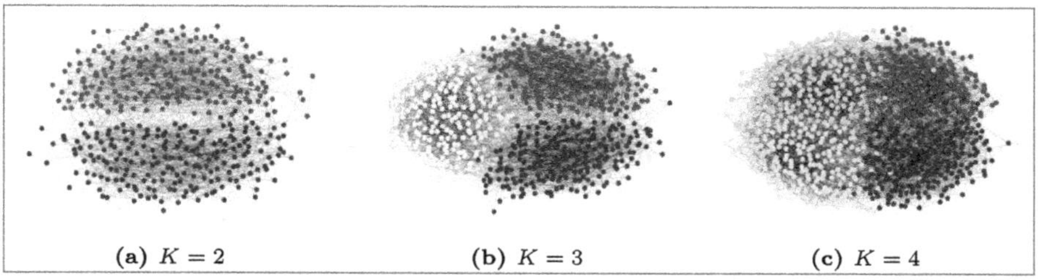

Figure 2.11. Different SBMs, with 200 nodes per community and connectivity probabilities $q_{kk} = 0.05$ and $q_{k\ell} = 0.005$ for $k \neq \ell$.

Proposition 2.4. *For $z \in [K]^n$ and $k \in [K]$, we denote $C_k^z = \{i \in [n] : z_i = k\}$ the community sets given by the node-labelling z. Let $(z, G) \sim \mathrm{SBM}(n, \pi, P)$. Then,*

$$\mathbb{P}(z) = \prod_{k=1}^{K} \pi_k^{|C_k^z|},$$

$$\mathbb{P}(G \mid z) = \prod_{1 \le i < j \le n} p_{z_i z_j}^{A_{ij}} (1 - p_{z_i z_j})^{1 - A_{ij}} \tag{2.4}$$

$$= \prod_{1 \le k \le \ell \le K} \left(p_{k\ell}\right)^{N_{k\ell}(1)} \left(1 - p_{k\ell}\right)^{N_{k\ell}(0)} \tag{2.5}$$

where $N_{k\ell}(a) = \sum_{1 \le i < j \le n} 1(A_{ij} = a) 1(z_i = k) 1(z_j = \ell)$ is the number of edges (if $a = 1$) or non-edge (if $a = 0$) between the communities k and ℓ.

Proof. By independence of the node community labels, we have

$$\mathbb{P}(z) = \prod_{i=1}^{n} \pi_{z_i} = \prod_{k=1}^{K} \pi_k^{|C_k^z|}.$$

Then, equation (2.4) is a consequence of Proposition 2.1. □

Remark 2.6. The adjacency matrix of $\mathrm{SBM}(n, \pi, P)$ can be seen as a block matrix, whose blocks are Erdős-Rényi graphs. This is in particular useful to efficiently simulate sparse SBM (see Algorithm 2, or the *networkX* or *iGraph* implementations).

Definition 2.7. We call *homogeneous* (or symmetric) SBM an SBM such that

$$p_{z_i z_j} = \begin{cases} p_{\mathrm{in}} & \text{if } \quad z_i = z_j \\ p_{\mathrm{out}} & \text{otherwise.} \end{cases}$$

Proposition 2.5. *Let (z, G) be sampled from a homogeneous SBM. Then,*

$$\begin{aligned}\mathbb{P}(G \mid z) \;=\; &\left(\frac{p_{\text{in}}}{1-p_{\text{in}}}\right)^{|E|}\left(1-p_{\text{in}}\right)^{\frac{n(n-1)}{2}} \times \\ &\times \left(\frac{1-p_{\text{out}}}{1-p_{\text{in}}}\right)^{\sum\limits_{1\le k<\ell\le n}|C_k^z|\cdot|C_\ell^z|}\left(\frac{p_{\text{out}}}{1-p_{\text{out}}}\frac{1-p_{\text{in}}}{p_{\text{in}}}\right)^{N_{\text{out}}^z}\end{aligned}$$

where $N_{\text{out}}^z = \sum_{1\le i<j\le n} 1(A_{ij}=1)1(z_i \neq z_j)$ is the number of inter-community edges.

Proof. From equation (2.5) we have

$$\mathbb{P}(G \mid z) \;=\; \prod_{1\le k\le \ell\le K}\left(p_{k\ell}\right)^{N_{k\ell}(1)}\left(1-p_{k\ell}\right)^{N_{k\ell}(0)}.$$

We notice that $N_{k\ell}(0)+N_{k\ell}(1) \;=\; \sum_{i<j} 1(z_i=k)1(z_j=\ell)$. Hence,

$$N_{k\ell}(0)+N_{k\ell}(1) \;=\; \begin{cases}\left|C_k^z\right|\cdot\left|C_\ell^z\right| & \text{if} \quad k\neq \ell,\\ \frac{|C_k^z|\cdot(|C_k^z|-1)}{2} & \text{otherwise,}\end{cases}$$

and

$$\begin{aligned}\mathbb{P}(G \mid z) \;=\; &\left(1-p_{\text{in}}\right)^{\sum\limits_{k=1}^{K}\frac{|C_k^z|\cdot(|C_k^z|-1)}{2}}\left(1-p_{\text{out}}\right)^{\sum\limits_{1\le k<\ell\le K}|C_k^z|\cdot|C_\ell^z|} \times \\ &\times\left(\frac{p_{\text{in}}}{1-p_{\text{in}}}\right)^{\sum_{k=1}^{K}N_{kk}(1)}\left(\frac{p_{\text{out}}}{1-p_{\text{out}}}\right)^{\sum\limits_{1\le k<\ell\le K}N_{k\ell}(1)}.\end{aligned}$$

Since $\sum\limits_{k=1}^{K}|C_k^z| = n$, then $\left(\sum\limits_{k=1}^{K}\left|C_k^Z\right|\right)^2 = n^2 - 2\sum\limits_{1\le k<\ell\le n}\left|C_k^z\right|\cdot\left|C_\ell^z\right|$, and

$$\begin{aligned}\mathbb{P}(G \mid z) \;=\; &\left(1-p_{\text{in}}\right)^{\frac{n(n-1)}{2}}\left(\frac{1-p_{\text{out}}}{1-p_{\text{in}}}\right)^{\sum\limits_{1\le k<\ell\le n}|C_k^z|\cdot|C_\ell^z|} \times \\ &\times\left(\frac{p_{\text{in}}}{1-p_{\text{in}}}\right)^{\sum_{k=1}^{K}N_{kk}(1)}\left(\frac{p_{\text{out}}}{1-p_{\text{out}}}\right)^{\sum\limits_{1\le k<\ell\le K}N_{k\ell}(1)}.\end{aligned}$$

Finally, since $N_{\text{out}}^z = \sum_{1\le k<\ell\le K} N_{k\ell}(1)$ and $|E| = \sum_{1\le k\le \ell\le K} N_{k\ell}(1)$ we have $\sum_{k=1}^{K} N_{kk}(1) = |E| - N_{\text{out}}^z$, and the proposition statement holds. □

Proposition 2.6. *Let* $(z, G) \sim \text{SBM}(n, \pi, P)$ *be a homogeneous SBM graph with uniform node labels, i.e.,* $\pi = \left(\frac{1}{K}, \ldots, \frac{1}{K}\right)$. *Then, the expected degree* $\bar{d}$ *of any node is*

$$\bar{d} = \left(\frac{n}{K} - 1\right) p_{\text{in}} + n\frac{K-1}{K} p_{\text{out}}.$$

Proof. It is similar to the proof of Proposition 2.2. A given node has $\frac{n}{K} - 1$ potential neighbors in its community, and $\frac{n}{K}(K-1)$ potential neighbors in the other communities. □

2.3.2 Degree-corrected Stochastic Block Model

The results established for Erdős-Rényi Model (giant component, connectivity, etc.) are also valid for the Stochastic Block Model. Moreover, the limitations mentioned for the Erdős-Rényi graphs also apply for SBMs, and in particular the limitation of the degree distribution. To introduce more heterogeneity in the degree distribution, Karrer and Newman, 2011 proposed the degree-corrected SBM. In this model, each node i has, in addition to its community label z_i, a degree parameter θ_i modelling its popularity (*i.e.*, the propensity of node i to make links). The formal definition is as follows.

Definition 2.8. Let n be the number of nodes and K be the number of communities, $\pi = (\pi_1, \ldots, \pi_K)$ be a probability vector, and P be a $K \times K$ symmetric matrix. Furthermore, let $\theta = (\theta_1, \ldots, \theta_n) \in \mathbb{R}_+^n$ be the vector of degree-correction parameters. The pair (z, G) is said to be drawn from the *Degree-corrected Stochastic Block Model* (DC-SBM) with parameters (n, π, P, θ) if

- $z = (z_1, \ldots, z_n) \in [K]^n$ is a random vector whose entries are independent and distributed according to π;
- conditionned on z, G is an undirected graph with n nodes, where nodes i and j are connected with probability $\min(\theta_i\theta_j P_{z_iz_j}; 1)$, independently of other node pairs.

In the following, we will always suppose that $\theta_i\theta_j P_{z_iz_j} < 1$ for every (i,j). From Definition 2.8, we notice that multiplying all the θ_i for i such that $z_i = k$ by a constant c, and dividing $P_{k\ell}$ by c if $k \neq \ell$ and P_{kk} by c^2 leads to the same model. Therefore, after sampling the community labelling z, we normalise the θ_i's, such that $\sum_i \theta_i 1(z_i = k) = n\pi_k$ where $n\pi_k$ is the expected number of nodes in block k. With this normalisation choice, we recover the SBM model if $\theta_i = 1$ for all i. Moreover, the parameter θ_i can be interpreted as the relative importance of node i in the graph. Another widely used normalisation consists in imposing $\sum_i \theta_i 1(z_i = k) = 1$.

Similarly to SBM, we define a homogeneous DC-SBM when the entries of matrix P take only two values: $P_{kk} = p_{\text{in}}$ and $P_{k\ell} = p_{\text{out}}$ for $k \neq \ell$.

Proposition 2.7. *Consider a pair* (z, G) *drawn under a homogeneous* DC-SBM (n, π, P, θ)*. Let* $i \in [n]$ *be a node in community* k*. The expected degree of* i *is given by*

$$\mathbb{E}\, d_i = \theta_i n \sum_{\ell=1}^{K} \pi_l P_{k\ell}.$$

Proof. Let A be the adjacency matrix of G. We have $d_i = \sum_{j=1}^{n} A_{ij}$, where the A_{ij} $(j = 1 \cdots n)$ are independent random variables distributed as $\operatorname{Ber}\left(\theta_i \theta_j P_{z_i z_j}\right)$. Hence, conditioning on z, gives

$$\mathbb{E}\,(d_i \,|\, z) = \sum_{j=1}^{n} \sum_{j=1}^{n} \theta_i \theta_j P_{z_i z_j} = \theta_i \sum_{\ell=1}^{K} \left(\sum_{j=1}^{n} 1(z_j = \ell) \theta_j \right) P_{k\ell}.$$

The result follows using the normalisation $\mathbb{E} \sum_{j=1}^{n} 1(z_j = \ell)\theta_j = n\pi_\ell$. □

To make some computations easier, it is sometimes convenient to define a *Poisson Degree-Corrected Block Model.* This refers to a random graph G with Poisson distributed edges. More precisely, $A_{ii} = 0$, and for $i \neq j$ we have

$$A_{ij} = A_{ji} \sim \operatorname{Poi}(\theta_i \theta_j \omega_{z_i z_j}), \tag{2.6}$$

where $\operatorname{Poi}(\lambda)$ denotes a Poisson random variable with parameter λ, whose probability mass function is given by

$$\mathbb{P}(X = k) = e^{-\lambda} \frac{\lambda^k}{k!}, \quad k = 0, 1, 2, \ldots.$$

The θ_i's are the degree-correction parameters and $w_{k\ell}$ is the edge density between blocks k and ℓ. Note that A_{ij} is then an integer-valued random variable. Similarly to the DC-SBM, we assume that for all $k \in [K]$, $\sum_i \theta_i 1(z_i = k) = n_k$. When all θ_i's are equal to one, we recover a Poisson version of the SBM, namely

$$A_{ij} = A_{ji} \sim \operatorname{Poi}\left(w_{z_i z_j}\right). \tag{2.7}$$

While these random graph models differ from the standard SBM and degree-corrected SBM by allowing integer-valued edges, we notice that when $n\omega_{k\ell} \ll 1$

the Poisson distribution is close to the Bernoulli distribution with the same parameter $\theta_i\theta_j\frac{\omega_{k\ell}}{n}$, hence making the two models similar in practice. The Poisson framework is interesting as it makes some computations easier. In particular, for the Poisson version we have

$$\begin{aligned}\mathbb{P}(A \mid z,\theta,\omega) &= \prod_{i<j}\frac{\left(\theta_i\theta_j\omega_{z_iz_j}\right)^{A_{ij}}}{A_{ij}!}\mathrm{e}^{-\theta_i\theta_j\omega_{z_iz_j}} \qquad (2.8)\\ &= \frac{\prod_i\theta_i^{d_i}}{\prod_{i<j}A_{ij}!}\prod_{1\le k\le K}\omega_{kk}^{m_{kk}}\mathrm{e}^{-\frac{n_k^2}{2}\omega_{kk}}\prod_{1\le k<\ell\le K}\omega_{k\ell}^{m_{k\ell}}\mathrm{e}^{-n_kn_\ell\omega_{k\ell}}\end{aligned}$$

where $n_k = \sum_i 1(z_i = k)$ is the number of nodes in block k and $m_{k\ell} = \sum_{i,j} A_{ij}1(z_i = k)1(z_j = \ell)$ denotes the number of edges going from community k to community ℓ (or twice the number if $k = \ell$).

2.3.3 Popularity Adjusted Block Model

While the DC-SBM allows us to accurately fit the degree distribution by enforcing the node degree parameters, it forces a popular node to be popular among all communities. Indeed, if θ_i is large, then node i will be expected to have a lot of friends in every communities. The *Popularity Adjusted Block Model* (PABM) bypasses this restriction, by allowing node popularity to vary across both nodes and communities.

Definition 2.9. Let n be the number of nodes, K be the number of communities, and $z \in [K]^n$ be a node-labelling vector. Let $\Lambda = (\lambda_{ik})_{i\in[n],k\in[K]} \in [0,1]^{n\times K}$. A graph $G = (V,E)$ is drawn from the Popularity Adjusted Block Model if $V = [n]$; the edges are formed independently; and

$$\mathbb{P}\left((ij) \in E\right) = \lambda_{iz_j}\lambda_{jz_i}.$$

In other words, λ_{ik} is the propensity of node i to form links with a node in community k.

Example 2.8. We recover the SBM by letting $\lambda_{ik} = \sqrt{P_{kl}}$ for every $i \in [n]$ and every $k \in [K]$.

Example 2.9. We recover the DC-SBM by letting $\lambda_{ik} = \theta_i\sqrt{P_{kl}}$ for every $i \in [n]$ and every $k \in [K]$.

2.3.4 Soft Geometric Block Model

Similarly to how the SBM extends the Erdős-Rényi model, the *Soft Geometric Block Model (SGBM)* extends the soft geometric random graphs or SERNs.

Definition 2.10. Let $(\mathcal{S}, d)$ be a metric space, and $(\gamma)_{1 \leq k,\ell \leq K} : \mathbb{R}_+ \to [0,1]$ be a set of connectivity functions, with $\gamma_{k\ell} = \gamma_{\ell k}$. We assign to each node a position $X_i \in \mathcal{S}$, and a community labelling $\sigma_i \in [K]$. Then,

$$\mathbb{P}(A \mid X, \sigma) = \prod_{i<j} \gamma_{\sigma_i \sigma_j} \left(d\left(X_i, X_j\right)\right)^{A_{ij}} \left(1 - \gamma_{\sigma_i \sigma_j}\left(d\left(X_i, X_j\right)\right)\right)^{1-A_{ij}}.$$

This model supposes that two nodes i, j are connected with probability that depends both on their position and their community assignment.

Example 2.10. We recover the SBM by further restraining $\gamma_{k\ell}(x) = q_{k\ell}$ to be constants (for all k, ℓ).

Example 2.11. The *Geometric Block Model (GBM)* restrains $\gamma_{k\ell}(x) = 1(x \leq r_{k\ell})$ for some parameters $r_{k\ell} \geq 0$.

Finally, the model is homogeneous if

$$\gamma_{k\ell} = \begin{cases} \gamma_{\text{in}} & \text{if } k = \ell, \\ \gamma_{\text{out}} & \text{otherwise.} \end{cases}$$

2.4 Exponential Random Graph Model

2.4.1 Definition and First Examples

Exponential Random Graph Model (ERGM) provides a convenient framework to explain the different *network statistics* observed in various networks. Examples of network statistics include the degree heterogeneity, the transitivity of relationships (friends of friends tend to be friends), the homophily (the propensity to link with nodes sharing the same attribute), the reciprocity of ties (in directed networks), etc.

Definition 2.11. Let n be the number of nodes, $\theta = (\theta_1, \ldots, \theta_q) \in \mathbb{R}^q$ be a vector of parameters, and $g = (g_1, \ldots, g_q)$ be a vector of network statistics. The adjacency matrix of an ERGM has the following probability distribution

$$\mathbb{P}(A \mid \theta) = \frac{\exp\left(\theta^T g(A)\right)}{\kappa(\theta)},$$

where $\kappa(\theta)$ is a normalisation constant.

Example 2.12. Consider the Bernoulli random graph model, where A_{ij} are independent, with $A_{ij} = A_{ji} \sim \mathrm{Ber}(p_{ij})$. Then

$$\mathbb{P}(A) = \prod_{i<j} p_{ij}^{A_{ij}} (1 - p_{ij})^{1-A_{ij}} = \frac{\exp\left(\sum_{i<j} \theta_{ij} A_{ij}\right)}{\kappa(\theta)} = \frac{\exp\left(\theta^T g(A)\right)}{\kappa(\theta)},$$

where $\theta_{i,j} = \log \frac{p_{ij}}{1-p_{ij}}$, $\kappa(\theta) = \left(\prod_{i<j}\left(1 - p_{ij}\right)\right)^{-1}$, $\theta = (\theta_{ij})_{1\le i<j\le N}$, and $g(A) = A_{ij}$ (we have $q = \binom{N}{2} = \frac{N(N-1)}{2}$ network statistics).

We notice in this Example that $\theta_{ij} = \log \frac{\mathbb{P}(A_{ij}=1)}{\mathbb{P}(A_{ij}=0)} = \operatorname{logit} \mathbb{P}\left(A_{ij} = 1\right)$ where $\operatorname{logit}(x) = \log \frac{x}{1-x}$. We will observe several of such relationships in the latter examples.

As the Erdős-Rényi graph, the SBM and DC-SBM are particular cases of the Bernoulli random graph, they can also be expressed as ERGM. For example, for an Erdős-Rényi random graph $\mathcal{G}_{n,p}$, we have $\theta_{ij} = \operatorname{logit}(p)$ is independent of i and j, and thus the previous example reduces to

$$\mathbb{P}(A) = \frac{\exp\left(\theta g(A)\right)}{\kappa(\theta)},$$

where $\theta = \log \frac{p}{1-p}$, $g(A) = \sum_{i<j} A_{ij} = |E|$ is the number of edges and $\kappa(\theta) = (1-p)^{-\frac{n(n-1)}{2}}$.

2.4.2 The p_1 Model

Now consider a directed graph A and let $X_{ij} = (A_{ij}, A_{ji})$. Assume that $(X_{ij})_{i<j}$ are independent, and define

$$\begin{aligned}
\mathbb{P}\left(X_{ij} = (1,1)\right) &= r_{ij}, \\
\mathbb{P}\left(X_{ij} = (1,0)\right) &= s_{ij}, \\
\mathbb{P}\left(X_{ij} = (0,0)\right) &= t_{ij}.
\end{aligned}$$

Note that $r_{ij} = r_{ji}$, $t_{ij} = t_{ji}$ and $r_{ij} + s_{ij} + s_{ji} + t_{ij} = 1$. Moreover,

$$\mathbb{P}(A) = \prod_{i<j} r_{ij}^{A_{ij}A_{ji}} \prod_{i\neq j} s_{ij}^{A_{ij}(1-A_{ij})} \prod_{i<j} t_{ij}^{(1-A_{ij})(1-A_{ji})}.$$

This can be re-expressed in an exponential form, as

$$\mathbb{P}(A) = \exp\left(\sum_{i<j} \rho_{ij} A_{ij} A_{ji} + \sum_{i \neq j} \mu_{ij} A_{ij}\right) \prod_{i<j} t_{ij},$$

where $\rho_{ij} = \log\left(\frac{r_{ij} t_{ij}}{s_{ij} s_{ji}}\right)$ and $\mu_{ij} = \log\left(\frac{s_{ij}}{t_{ij}}\right)$. We notice that

$$\mu_{ij} = \log\left(\frac{\mathbb{P}(A_{ij} = 1 \mid A_{ji} = 0)}{\mathbb{P}(A_{ij} = 0 \mid A_{ji} = 0)}\right) = \operatorname{logit}\left(\mathbb{P}\left(A_{ij} = 1 \mid A_{ji} = 0\right)\right)$$

measures the probability of an asymmetric link between i and j. Similarly,

$$\begin{aligned} \rho_{ij} &= \log\left(\frac{\mathbb{P}(A_{ij} = 1 \mid A_{ji} = 1)}{\mathbb{P}(A_{ij} = 0 \mid A_{ji} = 1)}\right) - \log\left(\frac{\mathbb{P}(A_{ij} = 1 \mid A_{ji} = 0)}{\mathbb{P}(A_{ij} = 0 \mid A_{ji} = 0)}\right) \\ &= \operatorname{logit}\left(\mathbb{P}\left(A_{ij} = 1 \mid A_{ji} = 1\right)\right) - \mu_{ij} \end{aligned}$$

is related to the probability that $A_{ij} = 1$ given that $A_{ji} = 1$, that is the *force of reciprocation* between i and j.

The *p_1-model* from (Holland and Leinhardt, 1981) further restricts $\rho_{ij} = \rho$ and $\mu_{ij} = \mu + \alpha_i + \beta_j$, so that

$$\mathbb{P}(A) = \frac{\exp\left(\rho R + \mu M + \sum_i \alpha_i A_{i+} + \sum_j \beta_j A_{+j}\right)}{\kappa(\rho, \mu, \alpha, \beta)}, \qquad (2.9)$$

where $A_{+i} = \sum_j A_{ji}$ denotes the in-degree of node i, and $A_{i+} = \sum_j A_{ij}$ the out-degree of node i, $M = \sum_{i,j} A_{ij}$ is the number of edges, and $R = \sum_{i,j} A_{ij} A_{ji}$ is the number of reciprocated edges. We can interpret equation (2.9) as follows:

- the parameter μ governs the density of (directed) edges. In particular, if $\rho = \alpha_i = \beta_j = 0$ while $\mu \neq 0$, then we recover a directed Erdős-Rényi random graph, with link-probability p such that $\mu = \operatorname{logit} p$;
- if α_i is large, then node i will tend to form an out-going edges. We can thus call α_i the *productivity* of node i;
- β_i refers to the *attractiveness* of node i, since a large β_i will push many nodes to form in-coming edges towards i;
- finally, the parameter ρ is the force of reciprocation of ties.

2.4.3 Relationship Between θ and the log-odds

We noticed in Example 2.12 that $\theta_{ij} = \operatorname{logit} \mathbb{P}\left(A_{ij} = 1\right)$, and similar relationships were drawn from the p_1-model. The following proposition generalizes it to any ERGM.

Proposition 2.8. *Consider an ERGM model as in Definition 2.11. Let* $A_{ij}^{+} = \{A \text{ with } A_{ij} = 1\}$ *be the graph with the edge* (i,j) *set to one,* $A_{ij}^{-} = \{A \text{ with } A_{ij} = 0\}$ *be the graph with the edge* (i,j) *set to zero, and* $A_{ij}^{c} = \{A_{uv} \text{ with } (u,v) \neq (i,j)\}$ *be the set of all the edges and non-edges except* A_{ij}*. Then, we have*

$$\operatorname{logit} \mathbb{P}\left(A_{ij} = 1 \,|\, A_{ij}^{c}\right) = \theta^{T}\left(g(A_{ij}^{+}) - g(A_{ij}^{-})\right).$$

Proof. Observe that

$$\begin{aligned}\mathbb{P}\left(A_{ij} = 1 \,|\, A_{ij}^{c}\right) &= \frac{\mathbb{P}\left(A_{ij}^{+}\right)}{\mathbb{P}\left(A_{ij}^{+}\right) + \mathbb{P}\left(A_{ij}^{-}\right)} \\ &= \frac{\exp\left(\theta^{T} g(A_{ij}^{+})\right)}{\exp\left(\theta^{T} g(A_{ij}^{+})\right) + \exp\left(\theta^{T} g(A_{ij}^{-})\right)}.\end{aligned}$$

Similarly,

$$\mathbb{P}\left(A_{ij} = 0 \,|\, A_{ij}^{c}\right) = \frac{\exp\left(\theta^{T} g(A_{ij}^{-})\right)}{\exp\left(\theta^{T} g(A_{ij}^{+})\right) + \exp\left(\theta^{T} g(A_{ij}^{-})\right)},$$

and therefore

$$\operatorname{logit} \mathbb{P}\left(A_{ij} = 1 \,|\, A_{ij}^{c}\right) = \theta^{T}\left[g(A_{ij}^{+}) - g(A_{ij}^{-})\right].$$

□

Further Notes

A nice additional reference for this chapter is Barabási, 2016 (an online and interactive version is available at: http://networksciencebook.com/), as well as (by order of relevance): Hofstad, 2016; Durrett, 2007; Chung and Lu, 2006. Finally, other classic books on random graphs (more focused on mathematical proofs) are Janson *et al.*, 2011; Bollobás, 2001.

Many random graph models exist. A model worth mentioning and not covered in this chapter is the small-world model (Watts and Strogatz, 1998). A complete review of the SBM is made in Abbe, 2018. For random geometric graphs, we refer the reader to Penrose, 2003.

A useful modification of the random geometric graph model with scale-free degree distribution is the hyperbolic geometric graph model (see e.g., Krioukov *et al.*, 2010).

DOI: 10.1561/9781638280514.ch3

Chapter 3

Network Centrality Indices

One natural question in network analysis is "which are the most important nodes in a network?" A node can be important in several aspects. For instance, in a social network, a node can be well-connected to many social groups or a node can facilitate the information flow in a network. In an information network, a node can provide links to important information sources or can be a reference node. In an infrastructure network, a node can be crucial for sustaining a good topological structure of a network.

The importance of nodes can be characterized by a real-valued function defined on network nodes. The values of the function indicate the degree of importance of the nodes and can be used for ranking purposes. Such functions are called *network centrality indices* or *network centrality measures*.[1] From the previous paragraph, it is already clear that different importance criteria potentially lead to very different definitions of centrality indices. Therefore, we first review various existing centrality indices and then discuss relations among them and several applications. Along with centrality indices for nodes, there are also centrality indices for network edges. Even

1. Even though the term *centrality measure* is more common in the literature, we prefer to use the term *centrality indices* to disambiguate from *probability measure*.

though our main focus will be on centrality indices for nodes, we shall mention some centrality indices for edges as well. Up to the present, many centrality indices have already been proposed and the list continues to grow. We try to overview the important distinctive cases.

3.1 Overview of Centrality Indices

In this section we divide the definitions of various centrality indices into groups. We admit that the proposed categorization is not the only possible and some centrality indices can be placed in more than one group. We shall try to mention possible re-classifications.

3.1.1 Distance Based Centrality Indices

Here we describe network centrality indices based on geodesic (shortest path) distances between the nodes.

Node degree. The simplest distance-based centrality index is the node degree, the number of immediate neighbours of a node. In the case of directed networks, we actually have *indegree* d_v^-, corresponding to the number of incoming edges, and *outdegree* d_v^+, corresponding to the number of outgoing edges. We note that the nodes with large indegree can be interpreted as "authorities" and the nodes with large outdegree can be interpreted as "hubs". In the context of bibliometrics, the indegree of an article is the number of the other articles citing this article, and the outdegree is the number of references present in the article. Naturally, an established authoritative article has many citations, and a survey article typically references many sources.

Closeness. Denote by $d(v, u)$ the length of a shortest path from node v to node u. Bavelas, 1950 has introduced the notion of *closeness centrality*. The closeness centrality index of node u can be defined by

$$\frac{n-1}{\sum_v d(v,u)}. \tag{3.1}$$

The closeness centrality is just a reciprocal of the average distance from the given node to all the other nodes. Originally, the closeness centrality was defined for undirected, connected networks. If the formal extension to the case of directed networks is quite straightforward, the absence of (strong) connectivity poses a problem.

Harmonic centrality. To overcome the problem of infinite path lengths in the case of disconnected or weakly connected networks, the notion of *harmonic centrality*

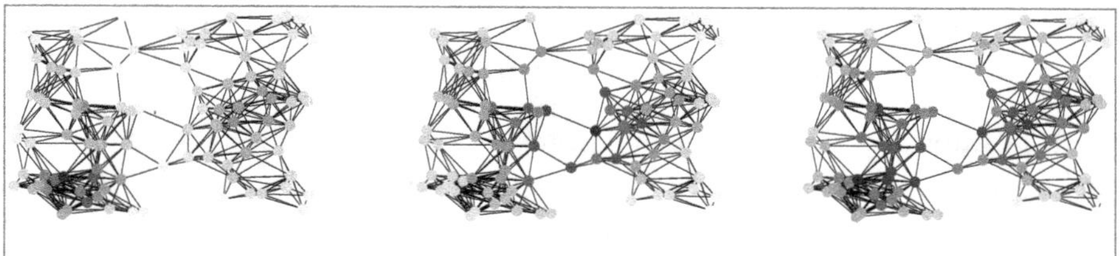

(a) Node degree (b) Closeness (c) Harmonic

Figure 3.1. Three distance based centrality indices (dark blue means high centrality).

was proposed. The idea of harmonic centrality is to swap the inversion and summation operations (also changing the normalisation), which results in

$$\frac{1}{n-1}\sum_{v:v\neq u}\frac{1}{d(v,u)}.$$

Thus, the harmonic centrality is the reciprocal of the harmonic mean distance. Thanks to the convention $\infty^{-1} = 0$, the harmonic mean naturally applies to disconnected or weakly connected networks. It appears that the notion of harmonic centrality was first proposed by Marchiori and Latora, 2000, even though several works independently proposed same notion or its variations, generalizations (Dekker, 2005; Cohen and Kaplan, 2007; Rochat, 2009; Pan and Saramäki, 2011).

Comparison. We calculate the above described three centrality indices on the same graph (see Figure 3.1). Of course, the node degree centrality is large only for high degree nodes, which in the chosen graph are located in the bottom left. On the contrary, closeness centrality gives importance to nodes at the junction of the two clusters. The harmonic centrality appears to mix both since nodes with large degrees as well as nodes located at the junction have large harmonic centrality values.

3.1.2 Spectral Centrality Indices

Spectral centrality indices are the indices that can be obtained as a solution of some eigenvalue problem

$$xM = \lambda x, \tag{3.2}$$

where x is a row-vector. A reason to operate with row vectors will be clear from the analysis that will follow.

Adjacency spectral centrality. This is one of the oldest centrality indices, whose application to scoring chess tournaments goes back to the end of the 19-th century (Landau, 1895). As the matrix M, we choose the graph adjacency matrix A and take as centrality indices the elements of the eigenvector associated with the largest positive eigenvalue. We note that such eigenvector is also called Perron-Frobenius eigenvector in the theory of non-negative matrices.

Random walk centrality or Seeley's index. Seeley, 1949 proposed to normalise the rows of the adjacency matrix by their sums. This implies that the reputation of a node is divided among the successors of that node. Thus, if we denote $P = D^{-1}A$, where D is the diagonal matrix of nodes' degrees, the *random walk centrality* is given as a solution of the following eigenvalue problem

$$\sigma = \sigma P. \tag{3.3}$$

There are two probabilistic interpretations of the elements of σ. The first interpretation says that σ_i is a long-term fraction of time a random walker on the graph spends at node i. Also, from the theory of Markov chains, we know that $E_i[T_i] = 1/\sigma_i$, where T_i is the return time to node i. Thus, by the second interpretation, the reciprocal of σ_i gives the expected return time to node i. Then, two further remarks are in order. Firstly, if the graph is undirected, the reversibility of the random walk, in this case, implies that Seeley's index becomes proportional to the node degree. Secondly, the original Seeley's index was defined only for strongly connected graphs. If a graph is not strongly connected, one can make various regularizations. One regularization will be described in the next paragraph.

PageRank. The creators of Google, Brin and Page, 1998 have proposed *PageRank centrality index* to rank web pages. PageRank models a web surfer behaviour by allowing a random walker to follow an out-going link with probability c and to restart from a uniformly random web page with the complementary probability $1 - c$. Thus, PageRank is the stationary distribution of the random walker, and hence it is a solution of the following system:

$$\pi = c\pi P + (1 - c)\nu, \tag{3.4}$$

where $P = D^{-1}A$ and ν is the uniform distribution. In fact, instead of the uniform distribution one can choose a distribution concentrated on some particular set of nodes. This results in *Personalized PageRank*, which allows one to measure centrality with respect to a certain group of nodes. Then, ν is referred to as the personalization distribution.

Note that using the normalisation condition $\pi \underline{1} = 1$, we can rewrite (3.4) as follows:

$$\pi = \pi(cP + (1 - c)\underline{1}\nu),$$

which explains why PageRank belongs to the family of spectral indices.

We can also rewrite equation (3.4) in the following way:

$$\pi [I - cP] = (1 - c)\nu,$$

which gives a useful explicit matrix expression for PageRank:

$$\pi = (1 - c)\nu [I - cP]^{-1}. \tag{3.5}$$

In particular, the above expression allows us to extend Seeley's index to non strongly connected networks. Consider first an intermediate situation when the network consists of m strongly connected components and each component is described by its own transition matrix $P^{(i)}$, $i = 1, \ldots, m$. Then, using formula (3.5) we can write

$$\pi = (1 - c)\left[\frac{n_1}{n}\frac{1}{n_1}\underline{1}^T \quad \cdots \quad \frac{n_m}{n}\frac{1}{n_m}\underline{1}^T\right]\begin{bmatrix}[I - cP^{(1)}]^{-1} & & \\ & \ddots & \\ & & [I - P^{(m)}]^{-1}\end{bmatrix}$$

$$= \left[\frac{n_1}{n}\pi^{(1)} \quad \cdots \quad \frac{n_m}{n}\pi^{(m)}\right],$$

where the vectors $\underline{1}$ are of appropriate dimensions and

$$\pi^{(i)} = (1 - c)\frac{1}{n_i}\underline{1}^T[I - cP^{(i)}]^{-1}$$

is PageRank of component i. Now we recall from the theory of Markov chains (see, e.g., Avrachenkov *et al.*, 2013a; Puterman, 2014) that the following asymptotic expansion takes place

$$[I - cP]^{-1} = \frac{1}{1 - c}\Pi + D + o(1 - c), \tag{3.6}$$

where Π is the ergodic projection and D is the deviation matrix, the quantities given by

$$\Pi = \lim_{T\to\infty} \frac{1}{T + 1}\sum_{t=0}^{T} P^t,$$

$$D = [I - P + \Pi]^{-1} - \Pi.$$

Now if a component is strongly connected, we have

$$\Pi^{(i)} = \underline{1}\sigma^{(i)},$$

where $\sigma^{(i)}$ is the stationary distribution of the random walker on component i, that is,

$$\sigma^{(i)} = \sigma^{(i)} P^{(i)}, \quad \sigma^{(i)} \underline{1} = 1.$$

Thus, it follows from (3.6) that

$$\pi^{(i)}(c) \to \sigma^{(i)} \quad \text{as} \quad c \to 1,$$

and a natural generalization of Seeley's index to the case of several stronly connected components is

$$\sigma = \left[\tfrac{n_1}{n} \sigma^{(1)} \quad \cdots \quad \tfrac{n_m}{n} \sigma^{(m)} \right], \tag{3.7}$$

where $\sigma^{(i)}$ is the stationary distribution or Seeley's index on component i. We see that in such generalization the relative importance of a component is proportional to its size, which appears to be quite fair. In particular, this generalization means that it is better to have a large "local" centrality in a large component.

The case of weakly connected components is treated in Avrachenkov *et al.*, 2008b.

PageRank with node-dependent restart probability. One natural generalization of PageRank is based on a random walk with restart that restarts with node-dependent probabilities. Specifically, let the random walk restart with probability $c(i)$ from node $i \in V$ with distribution ν. For convenience, define by C a diagonal matrix with $c(i)$ placed on its diagonal in the appropriate position. Then, the random walk with node-dependent restart can be described by the following transition probability matrix:

$$\tilde{P} = CD^{-1}A + (I - C)\underline{1}\nu. \tag{3.8}$$

Avrachenkov *et al.*, 2014a proposed two generalizations of the Personalized PageRank with node-dependent restart:

(i) The *Occupation-Time Personalized PageRank (OT-PPR)* is given by

$$\pi_j(\nu) = \lim_{t\to\infty} P_\nu[X_t = j]. \tag{3.9}$$

By the fact that $\pi(\nu)$ is the stationary distribution of the Markov chain, we can interpret $\pi_j(\nu)$ as a long-run frequency of visits to node j, *i.e.*,

$$\pi_j(\nu) = \lim_{t\to\infty} \frac{1}{t} \sum_{s=1}^{t} 1\{X_s = j\}.$$

(ii) The *Location-of-Restart Personalized PageRank (LR-PPR)* is given by

$$\rho_j(\nu) = \lim_{t\to\infty} P_\nu[X_t = j \text{ just before restart}]$$
$$= \lim_{t\to\infty} P_\nu[X_t = j \mid \text{restart at time } t+1]. \tag{3.10}$$

We can interpret $\rho_j(\nu)$ as a long-run frequency of visits to node j which are followed immediately by a restart, *i.e.*,

$$\rho_j(\nu) = \lim_{t\to\infty} \frac{1}{N_t} \sum_{s=1}^{t} 1\{X_t = j, X_{t+1} \text{ restarts}\},$$

where N_t denotes the number of restarts up to time t.

In Avrachenkov *et al.*, 2014a the following explicit matrix formulae were given for the Occupation-Time Personalized PageRank

$$\pi(\nu) = \frac{1}{\nu[I - CP]^{-1}\underline{1}}\nu[I - CP]^{-1}, \tag{3.11}$$

with $P = D^{-1}A$, and for the Location-of-Restart Personalized PageRank

$$\rho(\nu) = \nu[I - CP]^{-1}[I - C]. \tag{3.12}$$

We see that the formula (3.11) is indeed a generalization of (3.5).

Denote for brevity $\pi_j(i) = \pi_j(e_i^T)$, where e_i is the ith vector of the standard basis, so that $\pi_j(i)$ denotes the importance of node j from the perspective of i. Similarly, $\pi_i(j)$ denotes the importance of node i from the perspective of j. There is a very useful relation between these "direct" and "reverse" OT-PPRs in the case of *undirected* graphs.

Theorem 3.1 (Avrachenkov *et al.*, 2014a). *When $A^T = A$ and $C > 0$, the following relation holds*

$$\frac{d_i}{c_i K_i(C)}\pi_j(i) = \frac{d_j}{c_j K_j(C)}\pi_i(j), \tag{3.13}$$

with

$$K_i(C) = \frac{1}{e_i^T[I - CP]^{-1}\underline{1}}. \tag{3.14}$$

Note that $K_i(A)$ can be interpreted as the reciprocal of the expected time between two consecutive restarts if the restart distribution is concentrated on node i, *i.e.*,

$$K_i(A)^{-1} = E_i[\#\text{ steps before restart}]. \tag{3.15}$$

Thus, given that $[I - CP]^{-1}$ is the fundamental matrix of the absorbing Markov chain, the expression (3.11) admits one more probabilistic interpretation of the OT-PPR in the form of renewal equation

$$\pi_j(\nu) = \frac{E_\nu\,[\#\text{ visits to } j \text{ before restart}]}{E_\nu\,[\#\text{ steps before restart}]}.$$

In particular, if $c_i = c, \forall i$ (the case of standard PageRank), we obtain the following simple relation between "direct" and "reverse" PPRs:

Corollary 3.1. *When* $A^T = A$ *and* $c_i = c, \forall i$, *the relation* (3.13) *reduces to*

$$d_i \pi_j(i) = d_j \pi_i(j). \tag{3.16}$$

Katz's index. Katz, 1953 has proposed a centrality index, which is a clear predecessor of PageRank. It is given by the formula

$$\kappa = \underline{1}^T \sum_{t=1}^{\infty} \beta^t A^t = \underline{1}^T([I - \beta A]^{-1} - I). \tag{3.17}$$

Note that the subtraction of the identity is not really needed and one often refers to the following quantity as Katz index:

$$\kappa = \underline{1}^T \sum_{0=1}^{\infty} \beta^t A^t = \underline{1}^T [I - \beta A]^{-1}. \tag{3.18}$$

In order that the both versions will be well-defined the discounting parameter β should not exceed the reciprocal of the Perron-Frobenius eigenvalue, $\lambda^{-1}(A)$.

The main difference with PageRank is that Katz centrality gives "full endorsement" to each neighbour node pointed by an out-going link. It was observed that this could be appropriate in some social networks where a reference from one social network member to another member carries a lot of importance.

Vigna, 2016 has noticed that using the theorem of Brauer, 1952 on the displacement of eigenvalues, Katz index can be expressed as a solution of the eigenvalue problem:

$$\kappa = \kappa\,(\beta\lambda(A)A + (1 - \beta\lambda(A))r\underline{1}^T),$$

where r is the right dominant eigenvector of A such that $\underline{1}^T r = \lambda(A)$. This justifies the classification of Katz index as a spectral centrality index.

HITS centrality index. HITS, introduced by Kleinberg, 1999, actually provides two centrality indices. The first centrality index ranks nodes as authorities and the second centrality index ranks nodes as hubs. Kleinberg, 1999 suggests that a good "authoritative" node is pointed by many good "hubs", and in turn, a good "hub" points to good "authoritative" nodes. This verbal statement can be represented by the following iterative process:

$$h^{(k+1)} = a^{(k)}A^T,$$

$$a^{(k+1)} = h^{(k+1)}A,$$

where A is the adjacency matrix. In the limit, the authority index is a solution of

$$a = aA^T A.$$

Hence, the index a is the left dominant eigenvector of $A^T A$. Or equivalently, a is the left singular eigenvector associated with the largest singular eigenvalue of A. Similarly, the index h is the right singular eigenvector associated with the largest singular eigenvalue of A. Note that the above definition is only valid for strongly connected graphs.

Comparison. Figure 3.2 presents four spectral centrality indices on the same graph. We observe that the adjacency spectral centrality heavily weights the nodes in the bottom left of the graph, which appears to be a region with several large degree nodes. The random walk centrality gives more importance to other nodes as well, with an emphasis on larger degree nodes and nodes in the bottom left, while PageRank index diminishes even further the importance of nodes in the bottom left. In fact, as expected for the undirected graphs, due to time-reversibility, the random walk centrality gives the same ranking as the node degree (compare Figure 3.2b with Figure 3.1a). Finally, Katz centrality index gives importance to only a few nodes located in the middle of the left component.

3.1.3 Hitting Time Based Centrality Indices

It is often the case that not only the shortest paths but also longer paths matter in the analysis of social networks. A typical example of such a case is rumour or information propagation in social networks. In fact, this phenomenon was already reflected in the definition of PageRank and Katz centrality indices where all the paths are taken into account but longer paths are discounted. One more measure of "proximity" in networks is given by mean first passage times (or mean hitting times) of a random walker. The mean first passage time from node i to node j is given by (see e.g., Aldous and Fill, 2002; Meyer, 2000):

$$E_i[T_j] = e_i^T [I - P_{-j}]^{-1}\underline{1}, \tag{3.19}$$

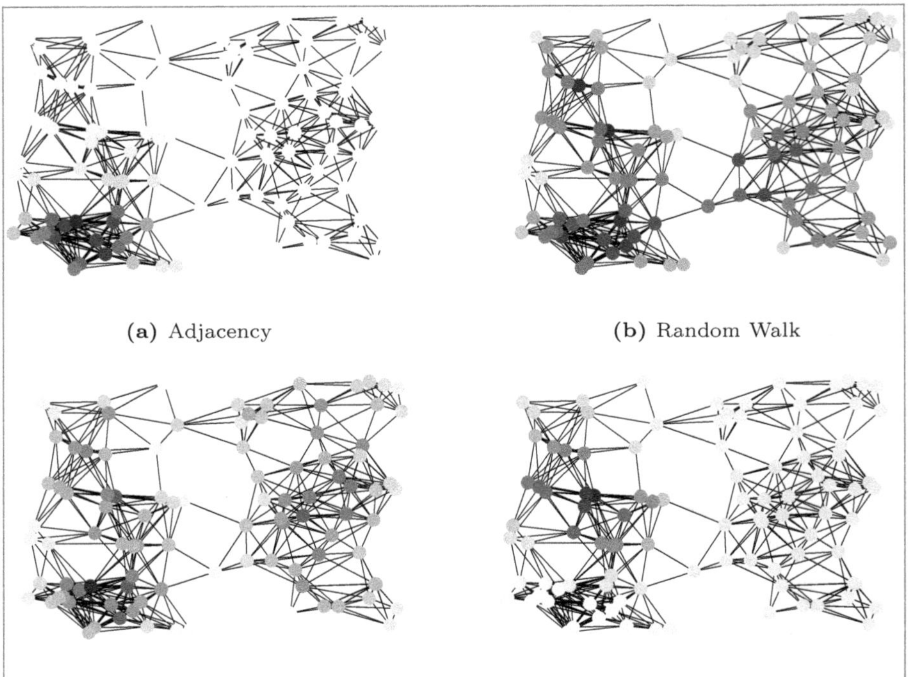

(a) Adjacency (b) Random Walk

(c) PageRank, $c = 0.85$ (d) Katz

Figure 3.2. Spectral centrality indices.

where e_i is the i-th vector of the standard basis and P_{-j} is the Taboo transition probability matrix obtained from P by deleting its j-th row and j-th column.

Now, in analogy with closeness centrality, see (3.1), we can define hitting time centrality as

$$h_j = \frac{n}{\sum_i e_i^T [I - P_{-j}]^{-1} \underline{1}} = \frac{n}{\underline{1}^T [I - P_{-j}]^{-1} \underline{1}}. \tag{3.20}$$

To the best of our knowledge, the expression (3.20) was proposed by White and Smyth, 2003. Note that in general $E_i[T_j] \neq E_j[T_i]$. Thus, one can also use the following alternative definition for hitting time centrality:

$$\tilde{h}_j = \frac{n}{e_j^T \sum_i [I - P_{-i}]^{-1} \underline{1}}. \tag{3.21}$$

It is known (see Chandra *et al.*, 1996; Aldous and Fill, 2002; Ellens *et al.*, 2011) that there is a connection between the effective resistance in a graph and hitting times:

$$r_{ij} = \frac{1}{2m}(E_i[T_j] + E_j[T_i]),$$

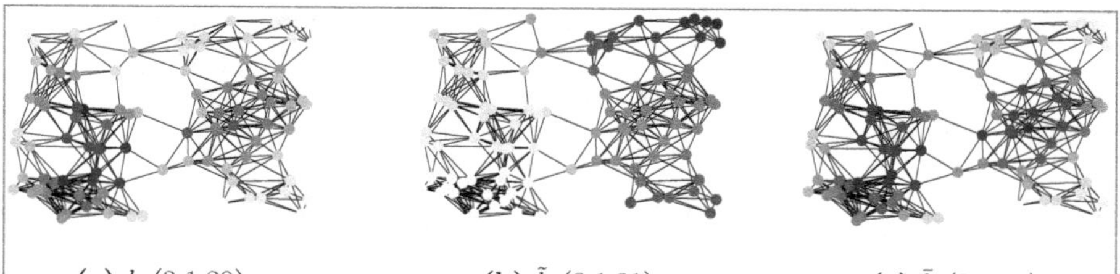

(a) h (3.1.20) (b) $\tilde{h}$ (3.1.21) (c) $\bar{h}$ (3.1.22)

Figure 3.3. Hitting time centrality indices.

where m is the number (total weight) of edges. Thus, a natural, symmetric version of hitting time centrality is given by

$$\bar{h}_j = \frac{1}{\sum_i r_{ij}} = \frac{2m}{\sum_i (E_i[T_j] + E_j[T_i])}. \tag{3.22}$$

One additional benefit of using effective resistances is that they actually define a metric on a graph.

Comparison. Figure 3.3 presents different hitting time centralities on the same graph. We observe that the centrality defined by (3.20) results in large weights on nodes in the bottom left and medium weights on nodes in the right component. On the contrary, the centrality defined by (3.21) gives small weights to nodes in the left component and large weights to nodes in the right component. Finally, the centrality defined by (3.22) results in large weights in nodes that are well connected, and small weights to more isolated nodes.

Extension to disconnected graph. The three versions of hitting time centrality (3.20), (3.21) and (3.22) are well-defined only for connected graphs. There are at least two approaches for extending this notion of centrality to disconnected or non strongly connected graphs. Firstly, one can just use the harmonic mean as was done in the case of the standard closeness centrality. Secondly, as was suggested in Hopcroft and Sheldon, 2008 and Avrachenkov *et al.*, 2018d, one can use the random walk with restart. Similarly to PageRank, let us consider the random walk with restart probability c. Then, the expected hitting time with restart from node i to node j is given by

$$E_i[T_j^c] = \frac{e_i^T [I - cP_{-j}]^{-1} \underline{1}}{1 - (1-c)\frac{1}{n}\underline{1}^T [I - cP_{-j}]^{-1} \underline{1}}.$$

The numerator of the above expression provides a more significant contribution than the denominator, especially when the parameter c is close to one. Thus, we suggest to use as a hitting time with restart centrality the following quantity:

$$h_j^c = \frac{n}{\underline{1}^T[I - cP_{-j}]^{-1}\underline{1}}. \tag{3.23}$$

Note that even though the network is strongly connected, the matrix $[I - P_{-j}]$ is often ill-conditioned and the introduction of the factor c helps to improve the condition number of the problem.

3.1.4 Betweenness Centrality Indices

A node in a social network can be viewed as important if that node contributes significantly to information flows or appears as a facilitator of communications.

Shortest path betweenness centrality. Freeman, 1977 introduced the betweenness centrality index based on shortest paths. Let σ_{st} be the number of shortest paths going from node s to node t, and let $\sigma_{st}(v)$ be the number of such shortest paths that pass through node v. Then, the shortest path betweenness centrality of node v is defined as follows:

$$\frac{1}{(n-1)(n-2)} \sum_{s,t:s,t \neq v} \frac{\sigma_{st}(v)}{\sigma_{st}}.$$

As was already mentioned, the information in social networks does not flow necessarily via shortest paths. Therefore, several researchers have extended the betweenness centrality to take into account longer paths.

Network flow betweenness centrality. In Freeman *et al.*, 1991, the authors suggest to use the concept of max-flow. This concept also allows to deal with weighted networks. Let w_{ij} be a weight of the link between nodes i and j (if there is no link, the weight is zero). Then, a flow from node s to node t is a mapping on the set of links such that the following two constraints are satisfied:

1. *Capacity constraint*: $\forall(i,j) \in E, f_{ij} \leq w_{ij}$;
2. *Conservation of flow*: $\forall v$ such that $v \neq s, t$:

$$\sum_{v:(u,v)\in E} f_{uv} = \sum_{v:(v,w)\in E} f_{vw}.$$

Then, the value of the flow f is given by

$$|f| = \sum_{v:(s,v)\in E} f_{s,v},$$

and the max-flow is the maximum flow which can go from s to t. Its value can be found by linear programming and the celebrated max-flow min-cut theorem says that the maximum flow is equal to the minimum capacity over all $s-t$ cuts.

Now, the network flow betweenness centrality of node v by Freeman *et al.*, 1991 is defined as follows:

$$\frac{\sum_{s,t:s,t\neq v} m_{st}(v)}{\sum_{s,t:s,t\neq v} m_{st}}, \tag{3.24}$$

where m_{st} is the value of max-flow from s to t and $m_{st}(v)$ is a part of such flow that passes through node v.

Current flow betweenness centralities. One more variant of betweenness centrality, which is based not only on shortest paths and uses the theory of electrical networks, was proposed by Brandes and Fleischer, 2005 and Newman, 2005a. Consider a weighted graph as an electrical network with conductances given by link weights. Suppose that a unit of current enters at node s (source) and leaves the network at node t (sink). Then, using Kirchhoff's current law and Ohm's law, we obtain the following linear system for the vector of potentials:

$$L\phi = b, \quad b_v = \begin{cases} 1, & v = s, \\ -1, & v = t, \\ 0, & \text{otherwise}, \end{cases} \tag{3.25}$$

where $L = D - A$ is the graph Laplacian. Since $L\underline{1} = 0$, the vector of potentials are determined up to an additive constant. Thus, without loss of generality, we can assume that the potential of the sink node is zero (this node is grounded). Then, the other potential values can be uniquely determined by (3.25). The throughput of node v is defined by

$$\tau_{st}(v) = \frac{1}{2}\left(-|b_v| + \sum_{w:(v,w)\in E} w_{vw}|\phi_v - \phi_w|\right),$$

and the current flow betweenness centrality is given by

$$\frac{1}{(n-1)(n-2)}\sum_{s,t}\tau_{st}(v). \tag{3.26}$$

Note that both the network flow betwenness (3.24) and the current flow betweenness (3.26) are well-defined only for strongly connected networks. In fact, the current flow betweenness is defined only for undirected networks. Furthermore, the system (3.25) is often ill-conditioned. To improve the conditioning of the system and to allow the application to not strongly connected networks, we can consider at least the following two regularizations. Firstly, as in the case of PageRank, we can regularize the system (3.25) as follows (Avrachenkov *et al.*, 2013b):

$$[D - \alpha A]\phi = b.$$

This modification has interpretations in terms of electrical network and random walks on graphs. In particular, this modification means that we multiply all the conductance by the factor α and ground each node with the conductance $(1-\alpha)d_v$. The other equations of the current flow centrality stay unchanged. The second regularization consists in adding a term to the Laplacian (Avrachenkov *et al.*, 2015):

$$[D - A + \beta I]\phi = b.$$

This has an interpretation that we ground all the nodes with conductance β, independent of node degree. Then, the above system has the following solution:

$$\phi = [I - D_2 P]^{-1} D_1 b,$$

with

$$D_1 = diag\left(\frac{1}{d_v + \beta}\right), \quad D_2 = diag\left(\frac{d_v}{d_v + \beta}\right).$$

Thus, the second regularization can be interpreted in terms of a random walk with non-uniform restart (see the paragraph about PageRank with non-uniform restart probabilities). Namely, the random walker restarts less frequently from high-degree nodes. As was observed in Avrachenkov *et al.*, 2013b and Avrachenkov *et al.*, 2015, both regularizations give similar rankings as the original current flow centrality but are much easier to calculate and to approximate. An advantage of the first regularization could be that the bias induced by high degree nodes is suppressed, whereas an advantage of the second regularization is in the fact that all the nodes are grounded in the same way and thus we need to perform averaging only over the source nodes.

We would like to mention that it is also very natural in the context of betweenness centralities to define centrality indices for edges. Specifically, for the shortest path edge betweenness centrality we count the number of shortest paths passing through an edge; and for the flow based edge betweenness centralities we calculate the flow passing through an edge under consideration. As we shall discuss later, the edge betweenness centralities are very useful for graph clustering.

Comparison. Figure 3.4 presents the different betweenness centralities calculated on the same graph. We observe that the shortest path betweenness centrality gives importance solely to nodes at the junction of the left and right clusters. Indeed, those nodes are of paramount importance in short paths since they are essential for joining a node in the left cluster to a node in the right one. Network flow gives

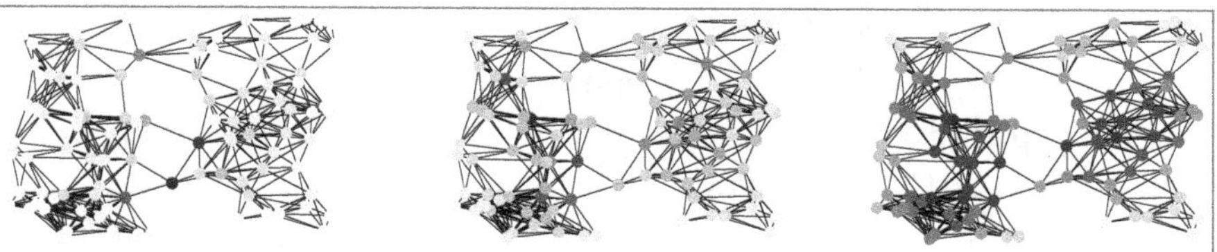

(a) Shortest path betweenness (b) Network flow (c) Current flow

Figure 3.4. Betweenness centrality indices.

more importance to nodes that are well connected and less importance to nodes that are more isolated. Finally, current flow enhances even further this, as nodes with low current flows are nodes in the extremity (top right or bottom right) with fewer connections to the rest of the graph.

3.1.5 Game Theory Based Centrality Indices

One more way to define a network centrality is based on cooperative game theory. This is actually quite a natural way to define network centrality since cooperative game theory provides means to estimate the importance of a node based on the node's contribution to network connectivity or network cohesiveness.

Let us recall that the basic quantity of cooperative game theory is a characteristic function $v(\cdot)$, which is defined on the subsets of nodes and satisfies the property $v(\emptyset) = 0$. Myerson, 1977 extended the concept of Shapley value, Shapley, 1953, to graph setting. *Meyrson-Shapley value* is a unique allocation, $Y_i(v, G)$, satisfying the following two axioms:

1. if S is a connected component of graph G, then the members of the coalition S ought to allocate to themselves the total value $v(S)$ available to them, *i.e.*,

$$\sum_{i \in S} Y_i(v, G) = v(S);$$

2. $\forall G, \forall i, j \in G$, both nodes i and j obtain equal payoffs after adding or deleting a link (i,j), *i.e.*,

$$Y_i(v, G) - Y_i(v, G - (i, j)) = Y_j(v, G) - Y_j(v, G - (i, j)).$$

The Myerson allocation can be computed by the Shapley formula

$$Y_i(v, G) = \sum_{S \subset V \setminus \{i\}} (v_G(S \cup \{i\}) - v_G(S)) \frac{s!(n - s - 1)!}{n!},$$

where $s = |S|$ and $v_G(\cdot)$ is the characteristic function defined additively with respect to connected components. However, in general, the computation by the above formula is very cumbersome. It appears that there is one natural choice for the characteristic function, which simplifies the computation of Myerson value. Inspired by path-discounting characteristic functions of Jackson and Wolinsky, 1996; Jackson, 2010, in Mazalov and Trukhina, 2014; Mazalov *et al.*, 2016 for trees and in Avrachenkov *et al.*, 2018a for general graphs, the following characteristic function was proposed. Let $\delta \in [0, 1]$ be a discount factor. Each link (or direct connection) gives to coalition S the value δ. Moreover, players obtain a value from indirect connections. Namely, each *simple* path of length 2 belonging to coalition S gives to this coalition the value δ^2, a simple path of length 3 gives to the coalition the value δ^3, etc. Thus, the coalition value can be expressed by the following formula

$$v(S) = a_1(G, S)\delta + a_2(G, S)\delta^2 + \cdots$$

where $a_k(G, S)$ is the number of simple paths of length k in coalition S. Recall that a simple path is a path with no repeated nodes. The use of simple paths is crucial. In Avrachenkov *et al.*, 2018a it was shown that this characteristic function leads to a manageable expression for the Myerson value:

$$Y_i(v, G) = \frac{a_1^{(i)}(G)}{2}\delta + \frac{a_2^{(i)}(G)}{3}\delta^2 + \cdots$$

where $a_k^{(i)}(G)$ is the number of simple paths of length k containing node i. The quantity $Y_i(v, G)$ as a centrality index combines some features of betweenness centrality with the path discounting as in PageRank and Katz centralities.

3.2 Axiomatic Comparison of Centrality Indices

As we have seen, there are many variants of centrality indices. Even inside the classes, such as distance-based indices or betweenness indices, there is a good number of variations. A big question (remaining largely open) is how to compare the centrality indices?

Of course, one can compare the indices numerically on some benchmark examples, which is a practically valid approach and we have seen examples of such comparisons in the first part of this chapter. One promising analytical approach is to propose a set of natural properties or axioms and test available centrality indices against those axioms. Such an approach was initially proposed by Boldi and Vigna, 2014. Let us describe it here.

The first two axioms of Boldi and Vigna, 2014 test centrality indices with respect to change of size and change of density. Two examples of strongly connected graphs with extreme densities are a cycle composed of links with the same direction and a clique with bi-directional links.

Size axiom. Consider the graph $G_{k,p}$ composed of a k-clique and a directed p-cycle. A centrality index satisfies the *size axiom*, if for every k there is $\bar{p}_k$ such that for all $p \geq \bar{p}_k$, the centrality of a node in the p-cycle is strictly larger than the centrality of a node in the k-clique. And, conversely, if for every p there is $\bar{k}_p$ such that for all $k \geq \bar{k}_p$ the centrality of a node in k-clique is strctly larger than the centrality of a node in the p-cycle.

Intuitively, the above axiom says that a node belonging to a very large but sparse community should be more important than a node belonging to a dense but small community.

However, the next axiom states that if communities are equal in size, a node belonging to a denser community should be more important.

Density axiom. Consider the graph $D_{k,p}$ composed of a k-clique and a directed p-cycle, which are connected by a bi-directional bridge, $x \leftrightarrow y$, with node x belonging to the clique and node y belonging to the cycle. A centrality index satisfies the *density axiom*, if for $k = p$ the centrality of x is strictly larger than the centrality of y.

Then, the third axiom says that it is natural that an immediate direct link always improves the index value of the node pointed by that link.

Score-monotonicity axiom. A centrality measure satisfies the *score-monotonicity axiom* if for every graph G and every pair of nodes x and y such that there is no link from x to y, when we add a link $x \to y$, the centrality of node y increases.

In the next Table 3.1, from Boldi and Vigna, 2014, we summarise the verification of the above axioms for most common centrality indices.

It is interesting to observe that, for the given selection of centrality indices, only harmonic centrality satisfies all the three axioms. This appears to be quite surprising taking into account how basic and simple are the requirements of the axioms. However, as will be demonstrated by application examples, we should not immediately discard the centrality indices that do not satisfy some axioms. They can be useful for tasks that are not described by those axioms.

3.3 Applications of Centrality Indices

3.3.1 Social, Bibliographic and Information Networks

Most definitions of centrality indices are originated in the domains of sociology and information networks. This is quite natural as centrality indices should indicate

Table 3.1. Axiom verification table, Boldi and Vigna, 2014.

Centrality	Size	Density	Score-monotonicity
Degree	only *k*	yes	yes
Closeness	no	no	no
Harmonic	**yes**	**yes**	**yes**
Betweenness	only *p*	no	no
Seeley	no	yes	no
Katz	only *k*	yes	yes
PageRank	no	yes	yes
HITS	only *k*	yes	no

which members of a social network are more important or powerful. Let us mention a few key contributions for centrality indices in sociology (the list is certainly not exhaustive): Bavelas, 1950; Bonacich, 1987; Bonacich and Lloyd, 2001; Borgatti, 2005; Brandes, 2008; Everett and Borgatti, 1999; Freeman, 1977; Freeman *et al.*, 1991; Friedkin, 1991; Hubbell, 1965; Katz, 1953; Newman, 2005a.

In bibliometrics, the citation count is simply the indegree centrality index for the citation network. (Citation networks have been introduced in the classical works by Solla Price, 1965, 1976.) Clearly, the citation count has its limitations. For instance, consider the case of an excellent original research article followed up by a comprehensive survey article. Over the years the survey article can accumulate more citations than the original research article, which could even be forgotten.

Chen *et al.*, 2007 have proposed to use PageRank to discover "scientific gems". They have ranked the publications in the Physical Review family of journals from 1893 to 2003 both by citation count and PageRank. Even though there appears to be a strong positive correlation between these two indices, there are articles with a very modest number of citations but with a very high PageRank score. These are often "scientific gems". For instance, a very important scientific technique or concept can be introduced in an article and then such concept can be named in honour of the inventors, and is used in many other articles but no specific reference is given anymore.

In recent work by Mariani *et al.*, 2016, the authors argue that PageRank identify well the established "scientific gems" but may miss new milestones. They proposed a rescaled PageRank, which takes into account the publication time.

The citation network is just one example of information networks. Other notable examples of information networks are the world wide web (see e.g., Brin and Page, 1998; Kleinberg, 1999; Hopcroft and Sheldon, 2008), the authorship network (the authors are nodes and the citations among the authors are the links), the co-authorship network (the articles are nodes and the links indicate if two papers were written by the same author[2]), journal citation network, etc. For instance, the authors of (Pinski and Narin, 1976; Bollen *et al.*, 2006; Bergstrom, 2007; Bergstrom *et al.*, 2008; González-Pereira *et al.*, 2010) use centrality indices, mostly spectral, to rank journals and the authors of (Fiala *et al.*, 2008; Ding *et al.*, 2009; Yan and Ding, 2009; Fiala, 2012; West *et al.*, 2013) use centrality indices to rank authors.

3.3.2 Semi-supervised Learning

Labelling data is a laborious and expensive process. Therefore, in many datasets the amount of labelled data is small and standard supervised machine learning methods either lead to a significant amount of errors or are not applicable at all. Fortunately, a graph-based semi-supervised learning method can help in such situations (Chapelle *et al.*, 2006).

The main idea behind graph-based semi-supervised learning is first to construct a graph on the data points, where a link between two data points indicate a strong relationship between these points. And then, one can use a similarity measure on the graph to assign unlabelled data points to the classes defined by the labelled data points.

One example of a similarity measure is given by the Personalized PageRank, see (Avrachenkov *et al.*, 2012). Suppose that class k is defined by a set of labelled points $\mathcal{L}_k$. And let ν_k be some (e.g., uniform) distribution with the support over the set $\mathcal{L}_k$. Then, we can define the similarity of data point u to class k by

$$\pi_u(k) = (1 - c)\nu_k[I - cP]^{-1} e_u,$$

with $P = D^{-1}A$. Thus, we attribute point u to class k, if

$$k = \arg\max_{k'} \pi_u(k').$$

An interested reader can find more examples of measures of node similarity in (Avrachenkov *et al.*, 2019). Many node similarity measures are related to centrality indices. We shall discuss semi-supervised learning in much more detail in Chapter 5.

2. Note that one can also consider a co-authorship network, where the authors are nodes and a link is present if two authors have co-authored an article. Erdős number network is one famous example of a co-authorship network.

3.3.3 Community Detection

The community detection problem is the problem of finding tight-knit groups of nodes in a network. We dedicate to this important topic the whole Chapter 4, but for now let us just point out some applications of centrality indices to the community detection problem.

Betweenness centrality indices are very efficient in solving the community detection problem. Specifically, in (Newman and Girvan, 2004) and (Newman, 2005a) the edges with the largest values of edge betweenness centrality are deleted and then the betweenness centrality is recomputed and then again the edges with the largest values of betweenness centrality are deleted, etc. This process will eventually lead to components that are disconnected between themselves.

The authors of (Avrachenkov *et al.*, 2008a) proposed first to find central nodes representing well communities with the help of PageRank and then as in the semi-supervised learning, to use Personalized PageRank to attribute nodes to communities.

Personalized PageRank and other centrality indices have also been applied to local graph clustering: Orponen and Schaeffer, 2005; Andersen *et al.*, 2006; Zhu *et al.*, 2013; Orecchia and Zhu, 2014; Gleich and Mahoney, 2014. This is clearly related to graph-based semi-supervised learning.

3.3.4 Further Applications

Historically, the first application of centrality indices was in sport and in particular in chess (Landau, 1895), which is followed by many other applications in this domain, just to name a few (Wei, 1952; Kendall, 1955; Keener, 1993; Callaghan *et al.*, 2007; Langville and Meyer, 2012).

Centrality indices play an important role in the analysis of network robustness (Albert *et al.*, 2000; Holme *et al.*, 2002; Ellens *et al.*, 2011; Rueda *et al.*, 2017; Ofori-Boateng *et al.*, 2021). In particular, Clemente and Cornaro, 2020 proposed new centrality indices characterizing the node and edges with respect to their effect on the vulnerability of a network.

Many recommender systems use centrality indices, in particular, centrality indices based on random walks: Fouss *et al.*, 2007; Gori *et al.*, 2007; Boldi *et al.*, 2008; Mei *et al.*, 2008; Fouss *et al.*, 2012; Davoodi *et al.*, 2013.

Centrality indices are also used in various NLP tasks such as semantic similarity (Sinha and Mihalcea, 2007), word sense disambiguation (Agirre and Soroa, 2009) and person name disambiguation (Smirnova *et al.*, 2010).

Further Notes

In addition to the work by Boldi and Vigna, 2014, there is a number of other works characterizing centrality indices by axiomatic approaches. Already Sabidussi, 1966 proposed several natural axioms to test centrality indices. In Altman and Tennenholtz, 2005 and Was and Skibski, 2018, the authors proposed axiomatization of Seeley's and PageRank centralities. Then, in Skibski and Sosnowska, 2018 the distance-based centralities were axiomatised.

In many applications, one needs to measure the centrality of a group of nodes rather than that of a single node. For instance, it may be required to evaluate an influence of a particular social group on the society or to evaluate the importance of a department within an organization. Several works, starting from Everett and Borgatti, 1999, proposed various variants of group centralities: Kolaczyk *et al.*, 2009; Veremyev *et al.*, 2017; Akgün and Tural, 2020. Let us emphasize that in the majority of cases it is not suitable to simply sum centrality values of individual nodes. It should come without surprise that the methods of cooperative game theory are very natural to define group centrality indices, see e.g., Michalak *et al.*, 2013; Szczepański *et al.*, 2016.

It is often important to find only Top-k central nodes of a network. This problem was investigated in Avrachenkov *et al.*, 2011, 2014c; Ostuni *et al.*, 2013; Avrachenkov *et al.*, 2014b; Yoshida, 2014; Borassi and Natale, 2019; Fan *et al.*, 2019; see also references therein.

We note that if in Katz centrality the geometric discounting is changed to the Poisson, factorial discounting, this gives Estrada's subgraph or communicability centrality (Estrada and Rodriguez-Velazquez, 2005; Estrada and Hatano, 2008). Furthermore, if now the adjacency matrix is replaced with the transition matrix of the random walk, this results in the heat kernel PageRank (Chung, 2007).

In the survey (Gleich, 2015) one can find an excellent extensive overview of various modifications and applications of PageRank.

DOI: 10.1561/9781638280514.ch4

Chapter 4

Community Detection in Networks

We saw in the introduction that the node set of many networks can be partitioned into several groups based on the node attributes or on the node's behaviour. For example, the members of the *karate-club network* split into two groups (Zachary, 1977), while the blogs of the *political blog network* are labelled as being either liberal or conservative. *Community detection* (also referred to as *community recovery* or *graph clustering*) consists in inferring the latent community structure based on the node's interactions.

Community detection is a delicate problem, as the notion of community is strictly speaking ill-defined. Indeed, although community structures are quite common in real networks, it is hard to properly define what is a community. Nonetheless, we can provide the following hints.

- **Definitions based on node similarity.** We can define as communities some groups of nodes that behave similarly to each other. For example, we could separate the nodes of a social network between influencers (people posting a lot of content and being followed by numerous users) and followers (peripheral nodes interacting mostly with influencers). To assess the similarity of two nodes, we can use a node similarity measure (such as Personalized PageRank or hitting time based centrality indices, see Chapter 3).

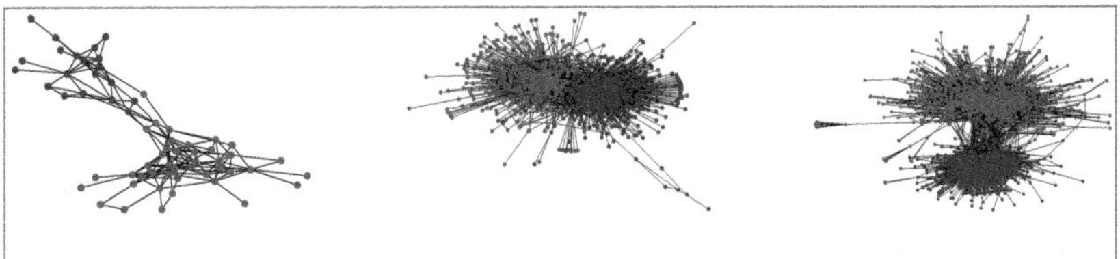

(a) Dolphin network. (b) Political blogs. (c) Live Journal.

Figure 4.1. Real networks with community structure.

- **Local definitions.** We can intuitively define a community as a set of nodes interacting a lot with each other. In that case, communities are groups of nodes that are densely connected within the groups but sparsely connected to the rest of the network.
- **Global definitions.** We can also evaluate the quality of a graph partition into disjoint communities using a quantity called modularity. This quantity compares the number of edges inside the community to the expected number of internal edges in a null model.

In addition to the above problem of choosing an appropriate definition of communities, we will see that community detection is often computationally difficult, and hence one needs to rely on approximation algorithms.

Real networks for which ground-truth communities are known are often used to compare and evaluate various community detection algorithms. We give below a non-exhaustive list of such networks. We plot some of them in Figure 4.1 and summarise in Table 4.1 some statistics of those networks. We also refer to the introduction, where these and some other networks were described in more detail.

- A selection of standard networks are available on Mark Newman personal webpage: http://www-personal.umich.edu/~mejn/netdata/, including the popular *Zachary Karate Club* (Zachary, 1977), an interaction network between dolphins (Lusseau *et al.*, 2003) and the *political blog* data set (Adamic and Glance, 2005).
- The Linqs webpage https://linqs.soe.ucsc.edu/data hosts some data set, including Cora, Citeseer, Pubmed, and WebKB (Lu and Getoor, 2003).
- The netset webpage https://netset.telecom-paris.fr/ hosts several data sets, including graphs with links between wikipedia articles.
- Finally, the *Stanford Large Network Dataset Collection* (https://snap.stanford.edu/data/) hosts a wide sample of larger networks.

Using synthetic networks is also common to assess the validity of community detection algorithms. A widely used random graph model with community structure is the Stochastic Block Model (SBM) and its degree-corrected variant (see Section 2.3).

Table 4.1. Selection of real data sets for community detection with ground-truth.

Category	Data Set	n	$\|E\|$	K	Features
Social networks	karate club	34	78	2	0
	dolphins	62	159	2	0
	LiveJournal top2	2766	24138	2	
Citation networks	cora	2485	5069	7	1433
	citeseer	2110	3668	6	3703
	DBLP top2	13326	34281	2	
Web networks	political blogs	1222	1671778	2	0
	wikischools	4403	100382	17	0
	wikivitals	10008	629521	11	0
Images	MNIST	70,000	–	10	784
	fashionMNIST	70,000	–	10	784
	CIFAR-10	60,000	–	10	
	CIFAR-100	60,000	–	100	

This chapter is structured as follows. We firstly present in Section 4.1 some cut-based methods, and their relaxation in the form of spectral clustering. Section 4.2 introduces modularity-based methods, and in particular the very efficient Louvain algorithm for modularity maximisation. The Bayesian framework for community detection is presented in Section 4.3. Moreover, in each section, we validate the proposed methods by performing numerical experiments and we discuss each method's limitations. Finally, we end the chapter with a theoretical analysis of the community detection problem in Section 4.4.

4.1 Cut-based Methods

In this section, we investigate the problem of separating the graph into $K \geq 2$ groups, such that inside a group the edge density is higher than between two different groups.

4.1.1 Graph Bisection

We consider a graph whose nodes are $\{1, \ldots, n\}$, and the adjacency matrix is $A = (a_{ij})_{1 \leq i,j \leq n}$. The graph is undirected but possibly weighted, so that $a_{ij} = a_{ji} \geq 0$. The degree d_i of a node $i \in V$ is defined as $\sum_{j=1}^{n} a_{ij}$.

Our goal is to partition the node set V into two subsets V_1, V_2 such that $V_1 \cap V_2 = \emptyset$ (non-overlapping communities) and $V_1 \cup V_2 = \{1, \ldots, n\}$. Note that $V_2 = V_1^c$, where V_1^c denotes $\{1, \ldots, n\} \backslash V_1$, the complementary set of V_1.

Definition 4.1. Given a set of nodes V_1 and an undirected graph represented by its adjacency matrix A, we denote by $\mathrm{Cut}(A, V_1)$ the total weight of the edges going from V_1 to its complement V_1^c. That is,

$$\mathrm{Cut}(A, V_1) = \sum_{i \in V_1, j \in V_1^c} a_{ij}.$$

At first glance, we might be tempted to solve

$$\widehat{V_1} = \underset{V_1 \subset [n]}{\arg\min}\ \mathrm{Cut}(A, V_1). \tag{4.1}$$

But, the trivial solutions of this minimisation problem are $\widehat{V_1} = V$ and $\widehat{V_1} = \emptyset$, which correspond to assigning every node to one cluster, and letting the second cluster empty! Moreover, even if we impose $V_1 \neq V$ and $V_1 \neq \emptyset$, we likely obtain a solution where almost all the nodes are in one cluster and only a few nodes in the other cluster.

Consequently, one can impose that the predicted sets V_1 and V_1^c should be roughly of the same size. To do so, one can penalize the imbalanced solutions. Firstly, let us impose the set V_1 and V_1^c to be exactly of the same size by restraining the minimisation problem over sets V_1 such that $|V_1| = \frac{n}{2}$. The new minimisation problem

$$\underset{V_1 \subset [n]\,:\,|V_1| = \frac{n}{2}}{\arg\min} \quad \mathrm{Cut}(A, V_1) \tag{4.2}$$

is called the *graph bisection problem*. Of course, in practice the two clusters are often of different size. This will be presented in the next section, along with a generalization to K clusters ($K \geq 2$).

But, even in this simple two cluster setting, another problem arises: the minimisation problem (4.2) is NP-hard (Wagner and Wagner, 1993; Garey *et al.*, 1974). Therefore, we have to rely on approximate methods. In the following, we propose a relaxation method based on the Laplacian. A similar method based on the adjacency matrix and a different method based on Semi-Definite Programming are presented in Section 4.1.3.

First relaxation method: Laplacian spectral clustering

Proposition 4.1. *For $V_1 \subset [n]$ such that $|V_1| = \frac{n}{2}$, define $z \in \{-1; 1\}^n$ the vector associated with the partition (V_1, V_1^c), that is $z_i = 1$ if $i \in V_1$ and $z_i = -1$ otherwise.*

We have

$$\underset{V_1 \subset [n]:\, |V_1| = \frac{n}{2}}{\arg\min} \ \mathrm{Cut}(A, V_1) = \underset{V_1 \subset [n]:\, |V_1| = \frac{n}{2}}{\arg\min} \ z^T L z.$$

Moreover, $z \perp 1_n$ *and* $\|z\|_2^2 = n$.

Proof. The facts that $z \perp 1_n$ and $\|z\|_2^2 = n$ are immediate from the constraint $|V_1| = n/2$. Moreover, we notice that

$$\left(z_i - z_j\right)^2 = \begin{cases} 1 & \text{if } i \in V_1, j \in V_1^c \text{ or } i \in V_1^c, j \in V_1 \\ 0 & \text{otherwise.} \end{cases}$$

Therefore,

$$\mathrm{Cut}(A, V_1) = \sum_{i \in V_1, j \in V_1^c} a_{ij} = \frac{1}{2} \sum_{i,j=1}^{n} a_{ij} \left(z_i - z_j\right)^2 = \frac{1}{4} z^T L z,$$

where the latter equality holds by Proposition A.10 from the background Section A.2. This ends the proof, as the factor $\frac{1}{4} > 0$ does not impact the minimisation problem. □

Minimisation problem (4.2) is therefore equivalent to

$$\hat{z} = \underset{\substack{z \in \{-1;1\}^n \\ \|z\|_2^2 = n \\ z \perp 1_n}}{\arg\min} \ z^T L z, \tag{4.3}$$

where the two associated clusters are simply $\widehat{V}_1 = \{i \in [n] : \hat{z}_i = 1\}$ and $\widehat{V}_1^c = \{i \in [n] : \hat{z}_i = -1\}$. A possible *continuous relaxation* of (4.3) is

$$\hat{x} = \underset{\substack{x \in \mathbf{R}^n \\ \|x\|_2^2 = n \\ x \perp 1_n}}{\arg\min} \ x^T L x.$$

By *relaxation*, we mean that we went from $z \in \{-1; 1\}^n$ to a real value vector $x \in \mathbb{R}^n$. This allows us to use standard calculus methods to solve the arg min problem (cf. Lemma 4.1). Once $\hat{x}$ is computed, we can cluster according to the sign of $\hat{x}_i$. This leads to the standard spectral clustering method (Algorithm 4). Nonetheless, there is in general no guarantee that the solution of the relaxed problem (4.3) should be equal to the true solution of the original problem (4.2). We refer to Section 4.4 for a more careful discussion.

Lemma 4.1. *Let $0 = \lambda_1 \leq \lambda_2 \leq \cdots \leq \lambda_n$ be the eigenvalues of L, and $v_1, \ldots, v_n$ the corresponding basis of orthogonal eigenvectors, normalized so that $\|v_i\|_2^2 = n$. We have*

$$\underset{\substack{x \in \mathbf{R}^n; \\ \|x\|_2^2 = n; \\ x \perp 1_n}}{\arg\min}\; x^T L x = v_2.$$

Proof. Indeed, $v_1 = 1_n$, and we conclude the proof by the Courant-Fischer Theorem (see Theorem A.13 in Appendix A.3.3). □

Algorithm 4: Standard Spectral Clustering – 2 clusters.

Input: graph standard Laplacian L.
Output: clustering assignment $\hat{z} \in \{1; 2\}^n$.
Spectral Step:

- let v_2 be the eigenvector of L associated to second smallest eigenvalue;
- for $i = 1 \ldots n$, let $\hat{z}_i = 1$ if $(v_2)_i > 0$, and $\hat{z}_i = 2$ otherwise.

Return: $\hat{z}$.

4.1.2 General Case: More Than Two Clusters

In this section, we extend the method of the preceding section to the general situation of $K \geq 2$ clusters of possibly different sizes.

Let $V_1, \ldots, V_K$ be a partition of V into K non-overlapping clusters, that is $V_1 \cup \cdots \cup V_K = V$ and $V_k \cap V_\ell = \emptyset$ for $k \neq \ell$. We highlighted in the preceding section the importance of penalizing the partitions of unbalanced cluster sizes in the Cut-minimisation problem. To measure the size of a cluster V_k, we define the two following metrics:

$$|V_k| = \sum_{i=1}^{n} 1(i \in V_k) \qquad \text{and} \qquad \mathrm{vol}\,(V_k) = \sum_{i \in V_k} d_i.$$

The quantity $|V_k|$ corresponds to the number of nodes belonging to the set V_k, while $\mathrm{vol}(V_k)$ is the volume of the set V_k, that is the sum of the degrees of nodes belonging to V_k. Instead of minimising directly the Cut, we will minimise one of these two quantities:

$$\mathrm{RatioCut}(A, V_1, \ldots, V_K) = \sum_{k=1}^{K} \frac{\mathrm{Cut}(A, V_k)}{|V_k|}, \tag{4.4}$$

$$\mathrm{NCut}(A, V_1, \ldots, V_K) = \sum_{k=1}^{K} \frac{\mathrm{Cut}(A, V_k)}{\mathrm{vol}(V_k)}. \tag{4.5}$$

The *Ratio-Cut* (resp., *Normalized-Cut* or *NCut*) corresponds to a Cut penalized with respect to the size (resp. volume) of the sets V_k: small sets bear a large penalty. Hence, we can expect that the solutions minimising the Ratio-Cut or the Normalized-Cut lead to clusters of balanced sizes.

As before, the minimisation of those quantities for all possible partitions $(V_1, \ldots, V_K)$ is NP-hard, and we will instead solve a relaxed version of the problem. Let us define the matrix $H = (h_{ik}) \in \mathbb{R}^{n \times K}$ by:

$$\forall\, i \in [n],\ \forall\, k \in [K]: \quad h_{ik} = \begin{cases} \dfrac{1}{\sqrt{|V_k|}}, & \text{if} \quad v_i \in V_k, \\ 0, & \text{otherwise.} \end{cases} \tag{4.6}$$

H is a matrix containing the K indicator vectors as columns, where the size of each set V_k is used as a normalisation term. Similarly, let us define $N = (n_{ik}) \in \mathbb{R}^{n \times K}$ as:

$$\forall\, i \in [n],\ \forall\, k \in [K]: \quad n_{ik} = \begin{cases} \dfrac{1}{\sqrt{\mathrm{vol}(V_k)}}, & \text{if} \quad v_i \in V_k, \\ 0, & \text{otherwise.} \end{cases} \tag{4.7}$$

Here we used the volume of each set V_k as a normalisation term. We have the following lemma.

Lemma 4.2. *The following holds:*

(i) $\mathrm{RatioCut}(A, V_1, \ldots, V_K) = \mathrm{Tr}\left(H^T L H\right)$;
(ii) $\mathrm{NCut}(A, V_1, \ldots, V_K) = \mathrm{Tr}\left(N^T L N\right)$;
(iii) $H^T H = I_K$ *and* $N^T D N = I_K$.

Proof. This lemma follows from the observations that

$$(H^T L H)_{kk} = H_{.k}^T L H_{.k} = \frac{\mathrm{Cut}(A, V_k)}{|V_k|},$$

where $H_{.k}$ denotes the column k of H, and

$$(N^T L N)_{kk} = N_{.k}^T L N_{.k} = \frac{\mathrm{Cut}(A, V_k)}{\mathrm{vol}(V_k)}.$$

Indeed,

$$\begin{aligned} H_{\cdot k}^T L H_{\cdot k} &= \frac{1}{2} \sum_{i,j} a_{ij} \left(h_{ik} - h_{jk}\right)^2 \\ &= \frac{1}{2} \left(\sum_{i \in V_k, j \notin V_k} a_{ij} + \sum_{i \notin V_k, j \in V_k} a_{ij} \right) \frac{1}{|V_k|} \\ &= \frac{1}{2} 2 \operatorname{Cut}(A, V_k) \frac{1}{|V_k|}. \end{aligned}$$

The second equality holds since $h_{ik} = h_{jk}$ if $(i \in V_k, j \in V_k)$ or $(i \notin V_k, j \notin V_k)$. The computations for $\left(N^T LN\right)_{kk}$ are similar. □

Therefore, minimising the RatioCut can be rewritten as:

$$\underset{(V_1,\ldots,V_k)}{\arg\min} \operatorname{Tr}\left(H^T L H\right), \tag{4.8}$$

where $L = D - A$, and H is defined in equation (4.6). Similarly, minimising NCut can be rewritten as:

$$\underset{(V_1,\ldots,V_k)}{\arg\min} \operatorname{Tr}\left(U^T \mathcal{L} U\right), \tag{4.9}$$

where $\mathcal{L} = D^{-1/2} L D^{-1/2}$, $U := D^{1/2} N$, and N is defined in equation (4.7).

The next step is to relax minimisation problems (4.8) and (4.9), by keeping only the constraints $H^T H = I_K$ and $U^T U = I_K$. The solution of these relaxed problems is given by the next proposition (we refer to the Proposition A.15 in Appendix A.3.3 for the proof).

Proposition 4.2. *Let $M \in \mathbb{R}^{n \times n}$ be a symmetric matrix. A solution of* $\arg\min \operatorname{Tr}\left(X^T M X\right)$ *where $X \in \mathbb{R}^{n \times K}$ is subject to $X^T X = I_K$ is given by the matrix $V \in \mathbb{R}^{n \times K}$ whose columns are the first K orthonormal eigenvectors of M.*

Once the relaxed problem is solved, we are left with a n-by-K matrix whose columns correspond to the K first eigenvectors of L (or $\mathcal{L}$). To reconvert this real-valued matrix to a discrete partition, a standard way is to consider the n rows of K (hence giving n data points in $\mathbb{R}^K$), and apply k-means algorithm on these n data points. More precisely, k-means consists in the following minimisation problem

$$\left(\widehat{Z}, \widehat{X}\right) = \underset{\substack{Z \in \mathcal{Z}_{n,K} \\ X \in \mathbb{R}^{K \times K}}}{\arg\min} \|ZX - V\|_F^2 \tag{4.10}$$

where $\mathcal{Z}_{n,K}$ denotes the space of *membership matrices*, that is $n \times K$ matrices with entries in $\{0, 1\}$ for which each row i has only one non-zero element. While solving the minimisation problem (4.10) is NP-hard, there exists (see Kumar *et al.*, 2004) a polynomial time procedure finding

$$\begin{aligned} &\left(\widehat{Z}, \widehat{X}\right) \in \mathcal{Z}_{n,K} \times \mathbb{R}^{K\times K} \\ \text{s.t.} \quad &\left\|\widehat{Z}\widehat{X} - V\right\|_F^2 \leq (1+\epsilon) \min_{\substack{Z\in\mathcal{Z}_{n,K} \\ X\in\mathbb{R}^{K\times K}}} \|ZX - V\|_F^2. \end{aligned} \tag{4.11}$$

Once $\widehat{Z}$ is found, we return the predicted clusters: node i is in cluster k if $\widehat{Z}_{ik} = 1$. We summarize this in Algorithm 5.

Algorithm 5: (Normalized) spectral clustering.

Input: graph Laplacian L (resp. normalized Laplacian $\mathcal{L}$), number of clusters K.

Output: predicted node labelling vector $\hat{z} \in [K]^n$.

Spectral Step:

- compute $v_1, \ldots, v_K$ the K orthonormal eigenvectors of L (resp. of $\mathcal{L}$) associated to the K smallest eigenvalues;
- let $V \in \mathbb{R}^{n\times K}$ be the matrix whose column k is v_k.

Clustering Step:

- let $\left(\widehat{Z}, \widehat{X}\right)$ be an $(1+\epsilon)$ approximate solution to the k-means problem (4.11);
- for every node $i = 1 \cdots n$, let $\hat{z}_i = k$ if $\widehat{Z}_{ik} = 1$.

Return: $\hat{z}$.

4.1.3 Semi-definite Programming

Similarly to the preceding section, we can also consider the problem of minimising

$$\text{Cut}(A, V_1, \ldots, V_K) = \sum_{k=1}^{K} \text{Cut}(A, V_k) \tag{4.12}$$

over the partitions $(V_1, \ldots, V_K)$ of V such that all the clusters V_k have equal size $|V|/K$. Similarly to what was done in the preceding section, we can show that *minimising* (4.12) is equivalent to *maximise*

$$\text{Tr}\left(X^T A X\right) \tag{4.13}$$

such that $X = (x_{ik})$ is a $n \times K$ matrix with

$$x_{ik} = \begin{cases} 1 & \text{if} \quad v_i \in V_k, \\ 0 & \text{otherwise.} \end{cases}$$

Maximising expression (4.13) leads to another Spectral Clustering method based on the adjacency matrix, where one look for the K eigenvectors associated to the K largest eigenvalues of A. We can also propose a different relaxation method. Indeed, from the relationship

$$\mathrm{Tr}\left(X^T A X\right) = \mathrm{Tr}\left(A X X^T\right), \tag{4.14}$$

it turns out that minimising (4.12) is equivalent to solving the following optimisation problem

$$\underset{\substack{Y \in \{0,1\}^{n \times n} \\ Y \succeq 0 \\ \mathrm{rank}(Y)=K \\ Y_{ii}=1 \\ Y 1_n = \frac{n}{K} 1_n}}{\arg\max} \; < A, Y > \tag{4.15}$$

where $< A, Y > = \mathrm{Tr}\left(A\, Y^T\right)$ denotes the usual matrix scalar product. The first four constraints in (4.15) force Y to be of the form XX^T while the last constraint forces the clusters to be of the same size.

A possible relaxation of optimisation problem (4.15) is the following semi-definite programming

$$\underset{\substack{Y \in \mathbb{R}^{n \times n} \\ Y \succeq 0 \\ Y_{ii} \leq 1 \\ Y 1_n = \frac{n}{K} 1_n}}{\arg\max} \; < A, Y > . \tag{4.16}$$

4.1.4 Discussion

Complexity of spectral clustering

Spectral methods require the computation of the eigenvectors, which has a worst-case complexity of $O\left(n^3\right)$. However, in practice when dealing with a sparse matrix whose eigenvalues are well separated, the complexity can be close to $O(Kn)$, where K is the number of eigenvectors needed (see *e.g.*, Demmel *et al.*, 2008).

Table 4.2. Performance of spectral clustering on real data sets.

Data Set	n	$\|E\|$	K	Accuracy
karate club	34	78	2	**94%**
dolphins	62	159	2	**98%**
political blogs	1,222	16,717	2	52%
DBLP-top2	13,326	34,281	2	55%
LiveJournal-top2	2766	24,138	2	**99%**
cora	2,485	5,069	7	37%
citeseer	2,110	3,668	6	59%
MNIST	70,000	784,186	10	63%

Performance of spectral clustering on real data sets

We first show in Table 4.2 the performance of spectral clustering, as it is implemented in the *scikit-learn* Python library[1]. This implementation uses the normalized Laplacian (and in practice, it has been observed that the normalized Laplacian outperforms the standard Laplacian).

We also show in Figure 4.2 the performance of normalized spectral clustering on the MNIST data set when we select two digits. We observe that most digit pairs are well predicted, but digit pairs $(4, 9)$, $(5, 8)$ and $(7, 9)$ are the hardest to distinguish, showing an accuracy of 0.53, 0.70 and 0.72, respectively. This highlights the intuitive fact that in those pairs the digits look similar.

Spectral methods and dangling trees

Let us analyse the failure of spectral clustering on the *political blogs* data set. Figure 4.3 shows the values of eigenvector components of $\mathcal{L}$ associated to the second and third smallest eigenvalues. We see that the entries of the second eigenvector are localised over a few nodes. Moreover, those nodes are associated to a dangling tree, and do not correspond to a meaningful community structure (see Figure 4.3c). On the contrary, the entries of the third eigenvector correspond to the correct community structure. In fact, using this eigenvector for clustering would lead to an accuracy of 95%.

Figure 4.3 shows that for this dataset, the good eigenvector for clustering is the third one, while the second eigenvector is concentrated around low degree nodes,

1. *sklearn.cluster.SpectralClustering*

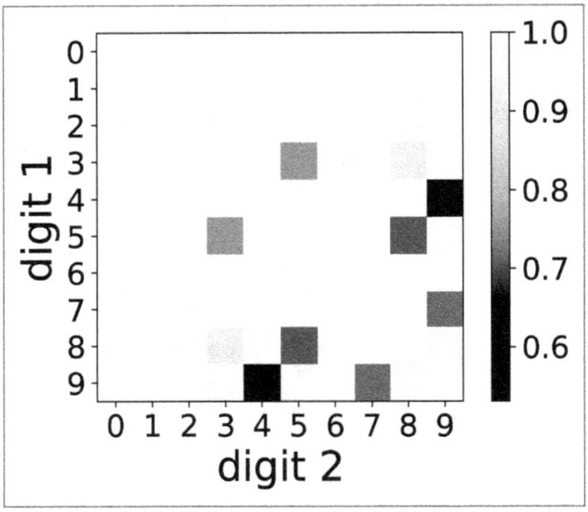

Figure 4.2. Accuracy of normalized Spectral Clustering on the MNIST data set restricted to two digits.

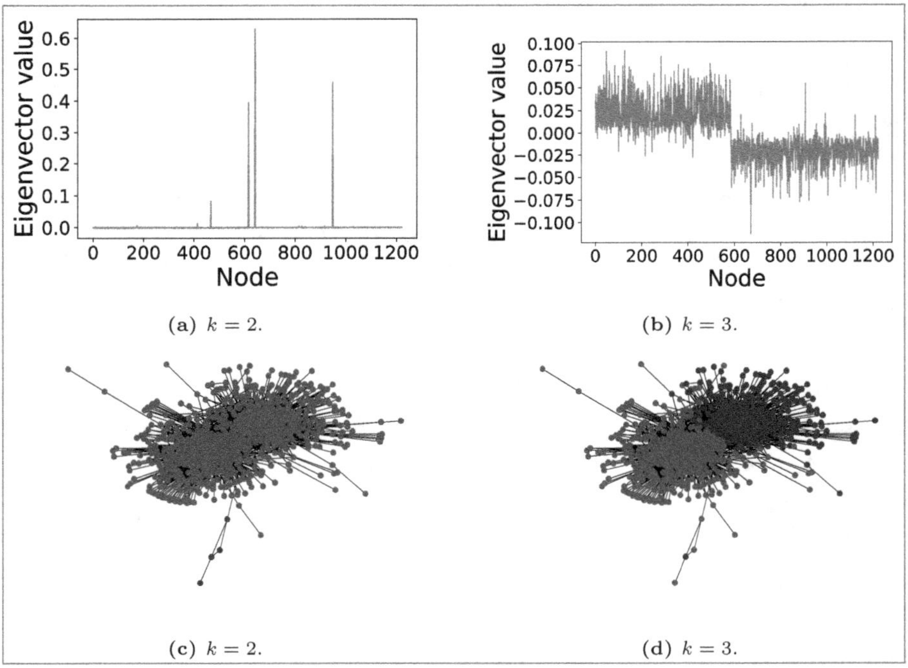

Figure 4.3. Analysis of the failure of spectral clustering on the *political blogs* data set. Top: values of the eigenvector components of $\mathcal{L}$ associated to the k-th smallest eigenvalue, for $k = 2$ and $k = 3$. Bottom: graph where the node colors correspond to the prediction made using the sign of the entries of the k-th eigenvector.

Table 4.3. Accuracy of regularized spectral clustering on real data sets, for different values of τ. Note that $\tau = 0$ corresponds to non-regularized spectral clustering.

	Regularized Spectral Clustering Accuracy		
Data Set	$\tau = 0$	$\tau = 1$	$\tau = \bar{d}$
political blogs	52%	**95%**	79% ($\bar{d} = 27.4$)
DBLP top2	55%	55%	55% ($\bar{d} = 5.1$)
cora	37%	**51%**	**52%** ($\bar{d} = 4.1$)
citeseer	**59%**	42%	32% ($\bar{d} = 3.5$)

forming a dangling tree.[2] Since this behavior results in partition of the graph into one large community with almost all the nodes and a small one with only a few nodes, it is easy to spot in practice. To solve this issue, one simple solution would be to look at higher order eigenvector. But, how to determine the correct eigenvector? Indeed, this might not always be an easy task. Firstly, it could happen that the correct eigenvector is in a lower position, say 5th or 7th, and localising it among noisy eigenvectors might be non trivial. Moreover, it is difficult to extend this reasoning for more than 2 clusters.

The *regularization technique* aims at solving this issue. It consists in performing spectral clustering on $\mathcal{L}_\tau := I - D_\tau^{-1/2} A_\tau D_\tau^{-1/2}$, where $A_\tau := A + \frac{\tau}{n} 1_n^T 1_n$ and D_τ is the associated transformed degree matrix. The matrix A_τ is a perturbed version of the initial adjacency matrix A, where we added an edge of weight $\frac{\tau}{n}$ between all nodes' pairs. This tends to bring back the dangling trees to the rest of the graph, hence restoring order in the eigenvectors (Zhang and Rohe, 2018). Moreover, Le *et al.*, 2017 showed that the regularized Laplacian $\mathcal{L}_\tau$ of Bernoulli random graphs is better concentrated around its expectation than the normalized Laplacian $\mathcal{L}$ (we develop further this point in Section 4.4.4, see in particular Theorem 4.9). The perturbation parameter τ is typically taken as $\tau = 1$ or $\tau = \bar{d}$ where $\bar{d}$ is the average degree of the graph. We compare in Table 4.3 the performance of standard spectral clustering with the regularized version.

Spectral methods and geometric data In many situations, nodes can have geometric attributes (for example a position in a metric space). As shown in Avrachenkov *et al.*, 2021a, this geometric structure handicaps cut-based clustering method. Indeed, in this case, the Fiedler vector might be associated to a geometric configuration, hence bearing no information about the latent community labelling. To avoid this pitfall, Avrachenkov *et al.*, 2021a proposed to look at

2. Note that if one were to use the standard Laplacian, we would also observe an analogous phenomenon, with noisy eigenvectors concentrated around high degree nodes.

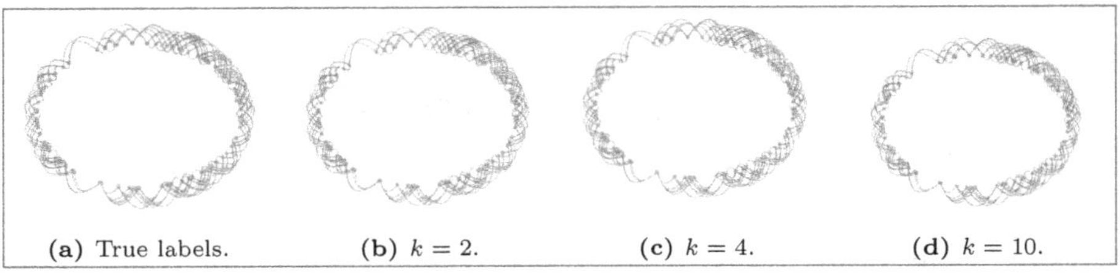

Figure 4.4. Analysis of the failure of spectral clustering on a Geometric Block Model, with 100 nodes and inter- and intra-distance cutoffs $r_{\text{in}} = 0.07$, $r_{\text{out}} = 0.02$.

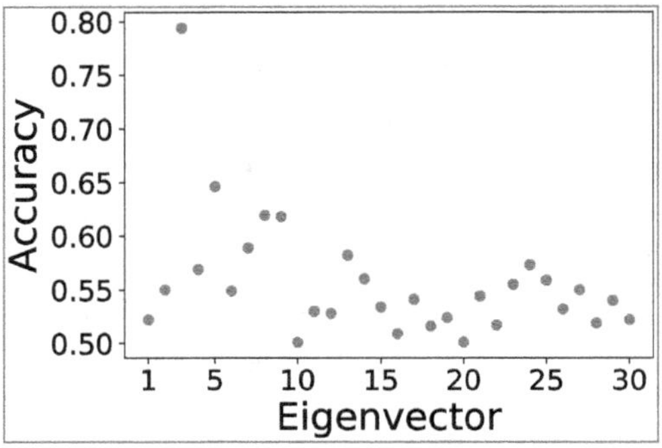

Figure 4.5. Accuracy obtained on weighted graph build using a subset of the MNIST data set ($n = 1000$ images representing digits 4 and 9) using the different eigenvectors of the normalized laplacian matrix $\mathcal{L}$. The eigenvector of index k corresponds to the eigenvector associated with the k-th smallest eigenvalue of $\mathcal{L}$.

higher order eigenvectors in order to recover the correct community memberships. Figure 4.4 highlights this situation. While the second and fourth eigenvectors give configurations based on the node location, recovering the node labels is better done with the 10-th eigenvector. The exact rang of the ideal eigenvector is then dependent on the model parameters, and we refer to Avrachenkov *et al.*, 2021a for a detailed analysis.

Let us also show that a higher order eigenvector can lead to a better clustering in real data sets with geometric components. We select 1000 images from MNIST, representing digits 4 and 9, and construct a k-nearest neighbors ($k = 8$) similarity graph with Gausian weights. The digits 4 and 9 form the hardest digit pair to distinguish. We plot in Figure 4.5 the accuracy obtained by spectral clustering as a function of the eigenvector order. We emphasis the fact that, unlinke the *politicl blog* data set, this is not an artifact due to dangling trees. We plot in Figure 4.6 the predicted clusters using the eigenvectors associated to the second and smallest eigenvalues of the graph's normalized Laplacian, and compare them with the true clusters. We notice that the predicted clusters are of balanced sizes. We also note that the NCut of the true labels is 3.8, while the NCut of the predicted labels

(a) True labels (b) Labels using v_2. (c) Labels using v_3.

Figure 4.6. Different clusterings on the same graph as in Figure 4.5. The colors on Figure 4.6(a) shows the true labels, while the colors on Figures 4.6(b) and 4.6(c) corresponds to the predicted labels using respectively the eigenvector associated to the second and third smallest eigenvalues of the normalized Laplacian.

associated with the prediction using the second (resp. third) eigenvector is 2.7 (resp. 3.7). Therefore for this graph, the correct labels do not correspond to the smallest normalized cut.

4.2 Modularity-based Methods

In this section, we will first define a quality function, called *modularity* (first introduced by Newman and Girvan, 2004), that aims to compare the density of links of our cluster assignment with the one we would obtain if the graph were build from a random null-model. By optimising the modularity over the space of all partitions, we identify groups of nodes that are more densely connected to each other than one would expect by random chance. As maximisation of modularity is NP-hard, we describe two common methods to do it approximately.

4.2.1 Definition

Definition 4.2. Given a vector $z \in [n]^n$ such that z_i denotes the community of node i, the *modularity* of z is defined by

$$\mathcal{M}(z) = \frac{1}{2|E|} \sum_{i,j} \left(A_{ij} - P_{ij}\right) 1 \left(z_i = z_j\right), \tag{4.17}$$

where $|E|$ is the number of edges and $P_{ij} = \dfrac{d_i d_j}{2|E|}$.

Remark 4.1. A few remarks are in order:

- We let the community labelling z take values in $[n]$, so that potentially we have n communities (hence every node can be alone in its community). Also, some communities can be empty.

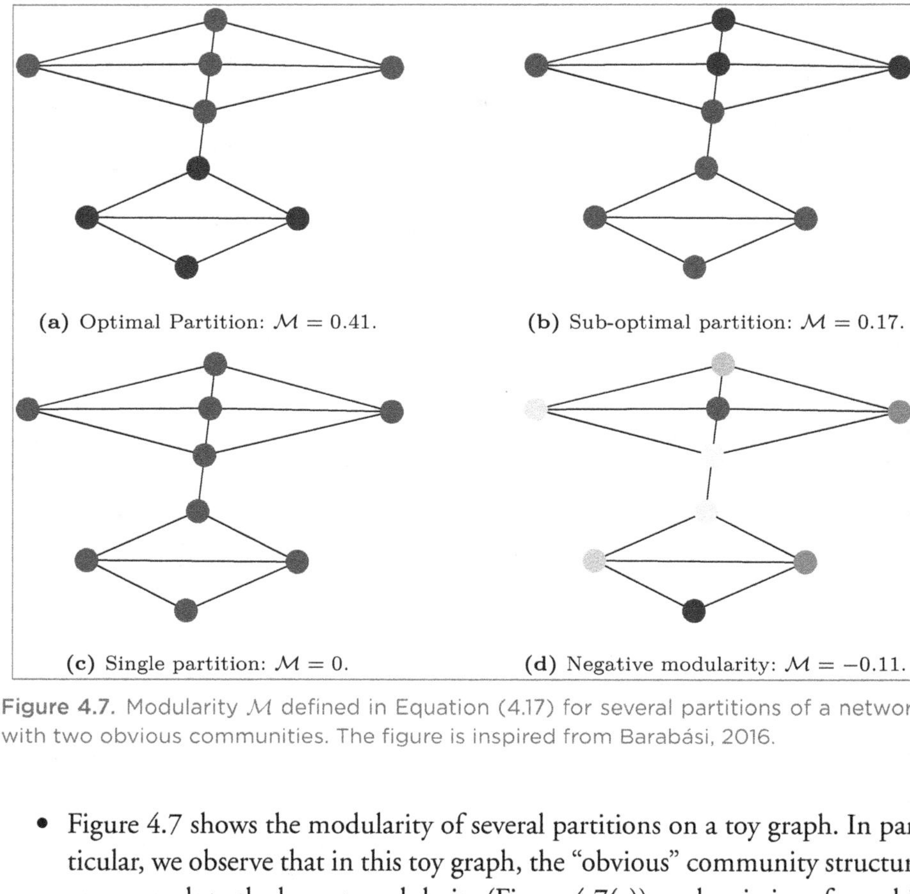

Figure 4.7. Modularity $\mathcal{M}$ defined in Equation (4.17) for several partitions of a network with two obvious communities. The figure is inspired from Barabási, 2016.

- Figure 4.7 shows the modularity of several partitions on a toy graph. In particular, we observe that in this toy graph, the "obvious" community structure corresponds to the largest modularity (Figure 4.7(a)), and variations from this partitions lead to smaller modularity (Figure 4.7(b)). Moreover:
  - if $z = 1_n$ (*i.e.*, the partition z assign all the nodes are in a single group), then $\mathcal{M}(z) = 0$ (Figure 4.7(c));
  - if the partition z assign all the nodes to be alone in their own community (*i.e.*, $z = (1, 2, \cdots, n)$), then $\mathcal{M}(z) \leq 0$ (Figure 4.7(d)).

  These two simple facts hold for any graph and are easy to establish.
- The factor $1/(2|E|)$ is a normalisation factor. In particular, showing that $-1 \leq \mathcal{M}(z) \leq 1$ for any graph and any node labelling vector z is straightforward[3].
- P_{ij} is the expected number of edges between i and j if the graph was drawn from a configuration model. Indeed, node i has d_i outgoing edges, and the probability that one of this edge goes to node j is $d_j/(2|E|)$, where $|E|$ is the

3. In fact, with some additional work, one can show that $-\frac{1}{2} \leq \mathcal{M}(z) \leq 1$ (Brandes *et al.*, 2007).

total number of edges in the network. In Section 4.4.1, we will further justify this choice by linking the modularity to the MAP estimator of a SBM.
- In practice, good values for modularity typically lie between 0.3 and 0.7. We refer to Table 4.4 for the modularity value of several networks with ground-truth community.
- Unfortunately, optimising the modularity is NP-complete (Brandes *et al.*, 2007).

Efficient computation of modularity

The following lemmas provide formulas to compute the modularity and to update the modularity, that will be useful for the algorithms presented in the next section. For a community labelling $z \in [n]^n$, we define the fraction of edges going from community k to community ℓ as

$$e_{k\ell}(z) = \frac{1}{2|E|} \sum_{i,j} A_{ij} 1(z_i = k) 1(z_j = \ell),$$

and the mass $m_k(C)$ of a community k as the sum of the degrees of the nodes in community k normalized by the total sum of node degrees

$$m_k(z) = \frac{1}{2|E|} \sum_{i=1}^{n} d_i 1\left(z_i = k\right).$$

Lemma 4.3. *The modularity of a community labelling is equal to*

$$\mathcal{M}(z) = \sum_{k=1}^{n} \left(e_{kk}(z) - (m_k(z))^2\right).$$

Proof. The proof is immediate, by writing

$$\mathcal{M}(C) = \frac{1}{2|E|} \sum_{k=1}^{n} \sum_{i,j} \left(A_{ij} - P_{ij}\right) 1(z_i = z_j = k)$$

and using the definitions of $e_{kk}(z)$ and $m_k(z)$. □

Lemma 4.4. *Let $z^{\text{old}} \in [n]^n$ be a community labelling, and define z^{new} as the labelling obtained by merging two communities k_1 and k_2:*

$$z_i^{\text{new}} = \begin{cases} k_1 & \textit{if } z_i^{\text{old}} = k_2 \\ z_i^{\text{old}} & \textit{otherwise.} \end{cases}$$

The resulting change of modularity is equal to

$$\mathcal{M}\left(z^{\text{new}}\right)-\mathcal{M}\left(z^{\text{old}}\right) = 2\left[e_{k_1k_2}\left(z^{\text{old}}\right)-m_{k_1}\left(z^{\text{old}}\right)m_{k_2}\left(z^{\text{old}}\right)\right].$$

Proof. For any $k \notin \{k_1, k_2\}$, $e_{kk}\left(z^{\text{old}}\right) = e_{kk}\left(z^{\text{new}}\right)$ and $m_k\left(z^{\text{new}}\right) = m_k\left(z^{\text{old}}\right)$. Moreover, since $\{i\colon z_i^{\text{new}} = k_2\} = \emptyset$, we have $e_{k_2k}(z^{\text{new}}) = 0$ and $m_{k_2}(z^{\text{new}}) = 0$. Therefore, using Lemma 4.3, the difference $\mathcal{M}\left(z^{\text{new}}\right)-\mathcal{M}\left(z^{\text{old}}\right)$ is equal to

$$e_{k1k_1}\left(z^{\text{new}}\right)-\left(m_{k_1}\left(z^{\text{new}}\right)\right)^2-\left(\sum_{k\in\{k_1,k_2\}} e_{kk}\left(z^{\text{old}}\right)-\left(m_k\left(z^{\text{old}}\right)\right)^2\right).$$

Since $\{i\colon z_i^{\text{new}} = k_1\} = \{i\colon z_i^{\text{old}} = k_1\} \cup \{i\colon z_i^{\text{old}} = k_2\}$, we have

$$e_{k_1k_1}\left(z^{\text{new}}\right) = e_{k_1k_1}\left(z^{\text{old}}\right)+2e_{k_1k_2}\left(z^{\text{old}}\right)+e_{k_2k_2}\left(z^{\text{old}}\right)$$

and

$$m_{k_1}\left(z^{\text{new}}\right) = m_{k_1}\left(z^{\text{old}}\right)+m_{k_2}\left(z^{\text{old}}\right),$$

which thus leads to the stated result. □

Lemma 4.5. *Let z^{new} and z^{old} be two community labellings that differ only for one node i. Let $z_i^{\text{old}} = a$ and $z_i^{\text{new}} = b$. Then the difference of modularity $\mathcal{M}\left(z^{\text{new}}\right)-\mathcal{M}\left(z^{\text{old}}\right)$ is equal to*

$$\left[e_{bb}\left(z^{\text{new}}\right)-m_b\left(z^{\text{new}}\right)^2\right]-\left[e_{aa}\left(z^{\text{old}}\right)-m_a\left(z^{\text{old}}\right)^2\right].$$

Proof. Since the only modified community are a and b, for any $k \notin \{a, b\}$, we have $e_{kk}(z^{\text{new}}) = e_{kk}(z^{\text{old}})$ and $m_k(z^{\text{new}}) = m_k(z^{\text{old}})$. The result then holds by Lemma 4.3. □

4.2.2 Greedy Algorithm

The first modularity maximisation algorithm, proposed by Newman, 2004, and reproduced here (Algorithm 6) iteratively joins pairs of communities if the move increases the partition's modularity. Some extension have been proposed (see for example Clauset *et al.*, 2004), but those have been outperformed by Louvain algorithm (Subsection 4.2.3).

Algorithm 6: Greedy algorithm for modularity maximisation.

Input: adjacency matrix A.
Output: node labelling $\hat{z} = (\hat{z}_1, \ldots, \hat{z}_n)$.
Initialize: assign each node to a community of its own, starting with n communities of single nodes (in other words, set $z_i = i$).
Update:

for *each community pair connected by at least one edge* **do**
- (i) Compute the modularity difference $\Delta\mathcal{M}$ obtained if we were to merge the two communities.
- (ii) Identify the community pair for which $\Delta\mathcal{M}$ is the largest and merge these two communities. (Modularity is always calculated for the full network, and $\Delta\mathcal{M}$ can be negative.)

Repeat the **Update step**, recording $\mathcal{M}$ at each step.
Stop when all nodes are merged into a single community.
Return: the partition $\hat{z}$ for which $\mathcal{M}$ is maximal.

Proposition 4.3. *The time complexity of Algorithm 6 is* $O\left(n\left(|E| + n\right)\right)$.

Proof. By Lemma 4.4, the computation $\Delta\mathcal{M}$ is done in constant time. At the initial update step, we have $|E|$ of such computations to do (and then at each update step, we have less than $|E|$, since we merge the communities). Then, after identifying the max $\Delta\mathcal{M}$ (which is done during the computations of all the $\Delta\mathcal{M}$), we need to recompute the adjacency matrix. This can take up to $O(n)$ operations. Finally, we need to do the update step $n - 1$ times. Hence, the overall time-complexity is of the order $n - 1$ times $|E| + n$. □

4.2.3 Louvain Algorithm

Algorithm 7 presents the *Louvain algorithm* invented by Blondel *et al.*, 2008. This method is called Louvain because the authors of the original paper were at that time based in Louvain University in Belgium.

Remark 4.2. Algorithm 7 requires to compute the change of modularity when one node is moved from one community to another. This can be done in constant time, as Lemma 4.5 shows.

Remark 4.3. The most time consuming pass of Algorithm 7 is the first pass, where we have $|E|$ changes of modularity to compute. The ensuing passes are faster, as they deal with much smaller graphs. Thus, a simple complexity estimate is $O(|E|)$, which is much better than the complexity of the greedy algorithm.

Algorithm 7: Louvain algorithm for fast modularity maximisation (Blondel *et al.*, 2008).

Input: adjacency matrix A.
Output: node labelling $\hat{z} = (\hat{z}_1, \ldots, \hat{z}_n)$.
Step I:

- assign each node to a community of its own, starting with n communities of single nodes (in other words, set $z_i = i$);
- for each node i, evaluate the gain in modularity if we place node i in the community of one of its neighbors j;
- move node i in the community for which the modularity gain is the largest, but only if this gain is positive. If no positive gain is found, i stays in its original community;
- apply this process to all nodes until no further improvement can be achieved. **In particular, a node can be moved several times.**

Step II: construct a network whose nodes are the communities identified in Step I, and where:

- the weight between two communities is the sum of the weights of the links between the nodes in the corresponding communities;
- the link between nodes of the same community lead to weighted self-loops.

1 **Step II** being completed, repeat **Step I** and then **Step II** (we call it a **pass**). Each pass decreases the number of communities. The passes are repeated until there are no more changes and a local maximum of the modularity is attained.
Return: $\hat{z}$.

4.2.4 Discussion

Unlike spectral methods, modularity based methods do not require the *a priori* knowledge of the number of blocks. Moreover, the speed difference observed in practice makes the greedy method (Algorithm 6) out of the competition. Furthermore, it has also been empirically observed that Louvain returns partitions with high modularity. Table 4.4 gives the performance of Louvain method on real data sets. In particular, we see that Louvain has the tendency to predict a high number of communities, but with a modularity higher than the ground truth partition.

We plot in Figures 4.8 and 4.9 the predicted communities by Louvain on the *karate club* and the *political blogs* datasets, respectively. By comparing to the ground truth, we observe that Louvain split the ground truth communities into smaller communities. It results in configurations with larger modularity that the ground truth ones (see Table 4.4), and the ground truth could be recovered almost perfectly by merging the small community predicted by Louvain into larger ones.

Table 4.4. Performance of Louvain algorithm on real data sets. K and $\mathcal{M}$ refer to the number of clusters and the modularity of the ground truth partition, while $\widehat{K}$ and $\widehat{\mathcal{M}}$ refer to the predicted number of cluster and the predicted modularity by Louvain algorithm.

Data Set Structure			**Ground Truth**		**Louvain**	
Name	n	$\|E\|$	K	$\mathcal{M}$	$\widehat{K}$	$\widehat{\mathcal{M}}$
karate club	34	78	2	0.36	4	**0.42**
dolphins	62	159	2	0.37	5	**0.52**
political blogs	1222	16717	2	0.41	11	**0.43**
DBLP top2	13326	34281	2	0.44	76	**0.86**
LiveJournal top2	2766	24138	2	0.38	21	**0.59**
cora	2485	5069	7	0.63	24	**0.80**
citeseer	2110	3668	6	0.52	37	**0.85**
wikivitals	10012	629527	11	0.31	8	**0.44**

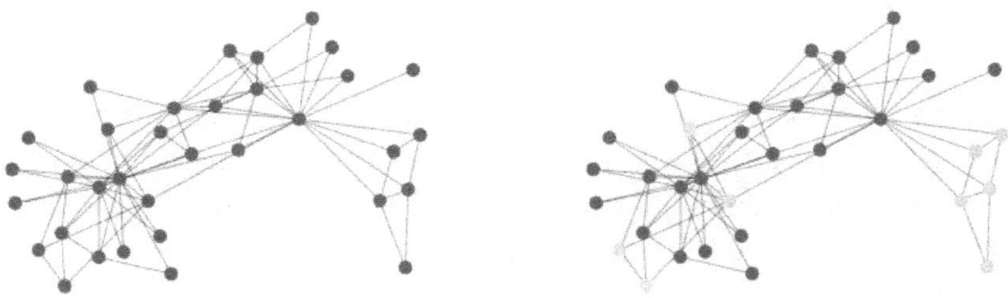

(a) True labels. (b) Labels predicted by Louvain.

Figure 4.8. Comparison of the ground-truth communities and the communities predicted by Louvain algorithm on the *karate club* dataset.

(a) True labels. (b) Labels predicted by Louvain.

Figure 4.9. Comparison of the ground-truth communities and the communities predicted by Louvain algorithm on the *political blogs* dataset.

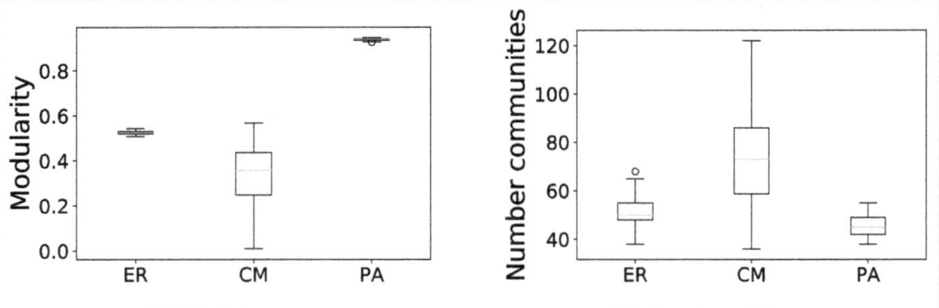

(a) Modularity. (b) Number of blocks.

Figure 4.10. Boxplot of the modularity and number of clusters obtained by Louvain algorithm on different random graphs without community structure. We computed 100 random graphs with $n = 1500$ nodes. ER refers to Erdős-Rényi model with $p = \frac{4}{n}$, CM to a configuration model with a Zipf law of parameter 2 for the degree distribution, and PA is the simple preferential attachment model as described in Section 2.2.2.

4.3 Bayesian Community Detection

4.3.1 An Over-fitting Issue?

Interpreting the result of any modularity maximisation algorithm should be done carefully. Indeed, partitions with high modularity can be found in random graph models without any community structure. Figure 4.10 shows the output (both modularity of the predicted partition and predicted number of clusters) of Louvain algorithm on Erdős-Rényi, configuration model and preferential attachment random graphs. The modularity found is high, especially in Erdős-Rényi and preferential attachment random graphs, albeit those graphs have by construction no community structure! Furthermore, partitions with high modularity are also found in configuration model, which is supposed to be the modularity's null-model. We emphasis that this is an intrinsic problem of modularity maximisation, and not a side-effect of the Louvain algorithm.

Cut-based methods are also prone to overfit. Figure 4.11 shows that using normalized spectral clustering on Erdős-Rényi random graphs leads to partitions whose cut represents between 15% and 30% of the total number of edges (depending on the number of clusters chosen). In other words, spectral clustering finds communities in the middle of pure randomness!

4.3.2 Principled Approach

To avoid the over-fitting issues and to find statistically significant communities in networks, we now explore a Bayesian approach. *Bayesian community detection* aims at determining which community labelling $z \in [n]^n$ is responsible of the network

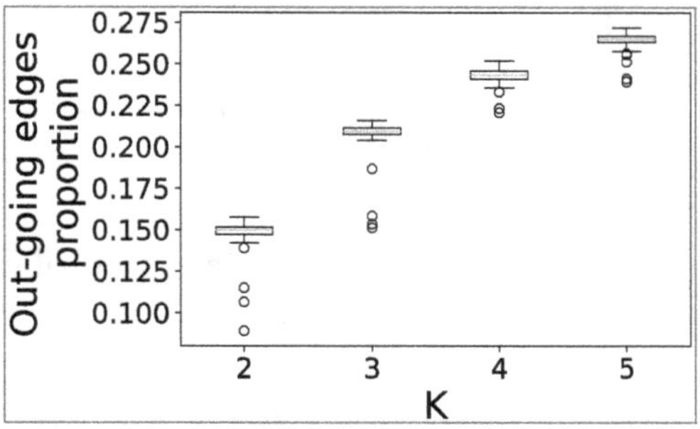

Figure 4.11. Boxplot of the proportion of out-going edges obtained by normalized spectral clustering for various K on Erdős-Rényi random graphs with $p = 0.01$.

A by maximising the *posterior distribution* $\mathbb{P}(z \mid A)$. Bayes' law gives

$$\mathbb{P}(z \mid A) = \frac{\mathbb{P}(A \mid z)\,\mathbb{P}(z)}{\mathbb{P}(A)}.$$

The quantity $\mathbb{P}(A)$ in the denominator is the *evidence*, that is the probability of the observed data, and does not depend on z.

The quantity $\mathbb{P}(A \mid z)$ is the *marginal likelihood*. We will make the assumption that the network was generated according to the Poisson version of the homogeneous DC-SBM (see Section 2.3.2). Therefore, $\mathbb{P}(A \mid z)$ is equal to

$$\int \mathbb{P}(A \mid z, \omega, \theta)\,\mathbb{P}(\omega \mid z)\,\mathbb{P}(\theta \mid z)\,d\omega\,d\theta. \tag{4.18}$$

In particular, $\mathbb{P}(A \mid z, \omega, \theta)$ is equal to[4] (see Proposition 2.8)

$$\prod_{1 \le k \le K} \omega_{kk}^{m_{kk}} \mathrm{e}^{-\frac{n_k^2}{2}\omega_{kk}} \prod_{1 \le k < \ell \le K} \omega_{k\ell}^{m_{k\ell}} \mathrm{e}^{-n_k n_\ell \omega_{k\ell}} \prod_i \theta_i^{d_i},$$

where $m_{k\ell} = \sum_{i<j} A_{ij} 1(z_i = k) 1(z_j = \ell)$. We select a uniform prior for θ that imposes the normalisation condition $\sum_i \theta_i 1(z_i = k) = n_k$ for all k. Hence

$$\mathbb{P}(\theta \mid z) = \prod_k (n_k - 1)!\,\delta\left(\sum_i \theta_i 1(z_i = k) - n_k\right).$$

Finally, we recall that for a continuous random variable $X \in [0, \infty)$ with constrained average $\bar{x}$, the maximum entropy distribution is the exponential distribution whose density is $f(x) = \mathrm{e}^{-x/\bar{x}}/\bar{x}$. Thus, we chose an exponential prior for $\omega_{k\ell}$,

4. up to a term $\prod_{i<j} A_{ij}!$ that does not depend on z.

such that

$$\mathbb{P}\left(\omega_{k\ell} \mid z\right) = \frac{\mathrm{e}^{-\omega_{k\ell}/\bar{\omega}}}{\bar{\omega}},$$

where $\bar{\omega} = 2|E|/n^2$ corresponds to the average edge probability in the network. Performing the integral over ω in Equation (4.18) leads to

$$\int \prod_k \frac{m_{kk}!}{\bar{\omega}\left(\frac{1}{\bar{\omega}} + \frac{n_k^2}{2}\right)^{m_{rr}+1}} \prod_{k<\ell} \frac{m_{k\ell}!}{\bar{\omega}\left(\frac{1}{\bar{\omega}} + n_k n_\ell\right)^{m_{k\ell}+1}} \prod_i \theta_i^{d_i} \mathbb{P}\left(\theta \mid z\right) d\theta,$$

where we used $\int_0^\infty \mathrm{e}^{-ax} x^b dx = \frac{b!}{a^{b+1}}$ for $a, b > 0$. To carry out the last integral over θ, we notice that for all k,

$$\prod_{i \in C_k} \int \theta_i^{d_i} \delta\left(\sum_{i \in C_k} \theta_i - n_k\right) d\theta_i = \frac{\prod_{i \in C_k} d_i!}{\left(\sum_{i \in C_k} d_i + 1\right)!}$$

where $C_k = \{i\colon z_i = k\}$. Therefore $\mathbb{P}\left(A \mid z\right)$ equals

$$\prod_k \frac{m_{kk}!}{\left(1 + \bar{\omega}\frac{n_k^2}{2}\right)^{m_{rr}+1}} \prod_{k<\ell} \frac{m_{k\ell}!}{\left(1 + \bar{\omega} n_k n_\ell\right)^{m_{k\ell}+1}}$$

$$\prod_k n_k^{v_k+1} \frac{(n_k - 1)!}{(n_k + v_k - 1)!} \frac{\bar{\omega}^{|E|} \prod_i d_i!}{\prod_{i<j} A_{ij}!}$$

where $v_k = \sum_i d_i 1(z_i = k)$ is the sum of the degrees of nodes in block k.

Let us now study the *prior distribution* $\mathbb{P}\left(z\right)$. In particular, the choice of prior should not make any *a priori* assumption on the number of (non-empty) groups and on the number of nodes in each groups (allowing groups of different sizes). Let

$$\mathbb{P}\left(z\right) = \mathbb{P}\left(z \mid \{n_k\}\right) \mathbb{P}\left(\{n_k\} \mid K\right) \mathbb{P}\left(K\right)$$

where K denotes the number of non-empty groups in σ, and n_k denotes the number of nodes in community k. We firstly have $\mathbb{P}\left(K\right) = \frac{1}{n}$ (the prior is agnostic about the number of blocks). Then, recalling that $\binom{n-1}{K-1}$ counts the number of ways to divide n nonzero counts into K nonempty bins, the probability that the K blocks have sizes $n_1, \ldots, n_K$ is $\mathbb{P}\left(\{n_k\} \mid K\right) = \frac{1}{\binom{n-1}{K-1}}$. Finally, given the randomly sampled block-sizes $\{n_k\}$, the partition is sampled with a uniform probability

$\mathbb{P}(\sigma \mid \{n_k\}) = \frac{\prod_r n_r!}{n!}\frac{1}{n}$. Therefore,

$$\mathbb{P}(z) = \frac{\prod_k n_k!}{n!} \cdot \frac{1}{\binom{n-1}{K-1}} \cdot \frac{1}{n}.$$

Using the expressions of the marginal likelihood and the prior leads to the maximisation of

$$\frac{1}{\binom{n-1}{K-1}} \prod_k \frac{m_{kk}!}{\left(1+\bar{\omega}\frac{n_k^2}{2}\right)^{m_{kk}+1}} \frac{n_k^{v_k}(n_k!)^2}{(n_k+v_k-1)!} \prod_{k<\ell} \frac{m_{k\ell}!}{(1+\bar{\omega} n_k n_\ell)^{m_{k\ell}+1}}$$

over all possible community labelling $z \in [n]^n$.

4.3.3 Markov Chain Monte Carlo Algorithm

While the above likelihood-based maximisation problem is hard, we can employ Markov Chain Monte Carlo (MCMC) importance sampling approach for finding a good approximate solution (Robert and Casella, 2013). We start from some initial labelling $z^{(0)}$. At each step, we propose a modification z' of the labelling $z^{(t)}$. This modification is accepted with probability $\min\left\{1, \frac{\mathbb{P}(z' \mid A)}{\mathbb{P}(z \mid A)} \frac{\mathbb{P}(z \mid z')}{\mathbb{P}(z' \mid z)}\right\}$. If the move is accepted, then $z^{(t+1)} = z'$, otherwise $z^{(t+1)} = z^{(t)}$. This acceptance probability is known as the *Metropolis-Hastings criterion*, and enforces the *detailed balance* (Metropolis *et al.*, 1953; Hastings, 1970a). Computing $\frac{\mathbb{P}(z' \mid A)}{\mathbb{P}(z \mid A)}$ has $O(d_i)$ time-complexity using the previous computations (in particular, we do not need to compute the evidence $\mathbb{P}(A)$ as it cancels out).

The simplest move proposal consists to select a node uniformly at random and choose its new community membership z_i' between the $K+1$ choice (the K existing groups plus the possibility to assign i to an empty group). This direct approach is inefficient, as the mixing time of the Markov Chain might be enormous. A better approach (Peixoto, 2014a, 2019) consists in choosing the new group membership z_i' according to

$$\mathbb{P}\left(z_i' = \ell \mid z\right) = \sum_k \mathbb{P}(k \mid i) \frac{e_{k\ell} + \epsilon}{e_k + \epsilon(K+1)}$$

where $\mathbb{P}(k \mid i) = \sum_j \frac{A_{ij} 1(z_j = k)}{d_i}$ is the fraction of neighbors of i belonging to group k, and $\epsilon > 0$ is a parameter enforcing ergodicity. We can interpret this probability as firstly choosing a node i uniformly at random, and sampling a neighbor j of i,

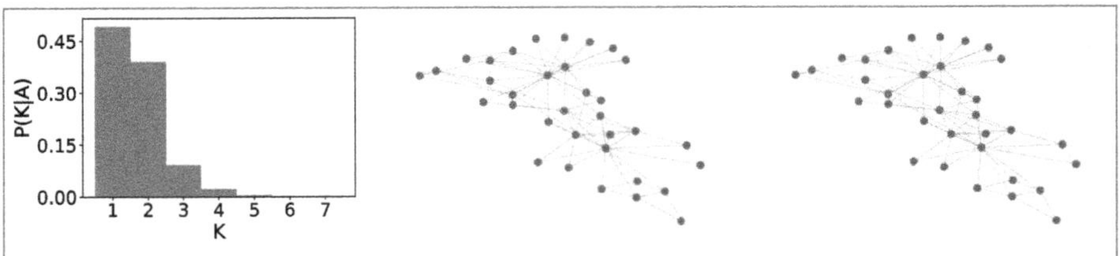

(a) Number of communities. **(b)** Community memberships. **(c)** Community memberships.

Figure 4.12. Marginal posterior probability of the number of groups (Figure 4.12(a)) for the karate-club network, under the assumption that the network is a realization of a degree-corrected SBM. Figures 4.12(b) and 4.12(c) show example of partitions obtained.

whose community label is $z_j^{(t)} = k$. Then,

(i) with probability $\frac{\epsilon}{e_k+\epsilon(K+1)}$ we choose a community label ℓ at random among the $K+1$ possibilities (it can be an empty group);
(ii) otherwise, we sample a group label ℓ with probability $\frac{e_{k\ell}}{e_k+\epsilon(K+1)}$.

This procedure can be performed in $O(d_i)$ time-complexity, provided that we keep track of the edges incidents from each group, which incurs $O(E)$ memory-complexity.

4.3.4 Numerical Results

The MCMC algorithm described in this section is implemented in the *graph-tool* library (Peixoto, 2014b), available at http://graph-tool.skewed.de.

We first analyze the performance of Bayesian clustering on synthetic networks. We generate DC-SBM graphs.

The MCMC procedure for the Bayesian framework gives access to the posterior distribution, instead of just finding its maximum. We can in particular obtain the marginal probabilities of group memberships of the network as well as marginal probability on the number of groups. In particular, we plot in Figure 4.12 the results obtained on the karate-club network. In particular, we observe on Figure 4.12(a) a large probability for the network to have one or two communities and the configurations with larger numbers of communities are much less likely. Recalling that the ground-truth corresponds to the situation after the feud between the main instructor and the club's president, we can interpret the large posterior on the one community case as the network before the feud, in which no communities were present at that time. When Bayesian clustering predicts two communities, we observe different configurations. Some predictions do align with the two communities observed after the feud (see Figure 4.12(b)), while other configurations tend to group the large degree nodes, 'influencers', together, and the low degree nodes, 'followers', forming the second community (see Figure 4.12(c)).

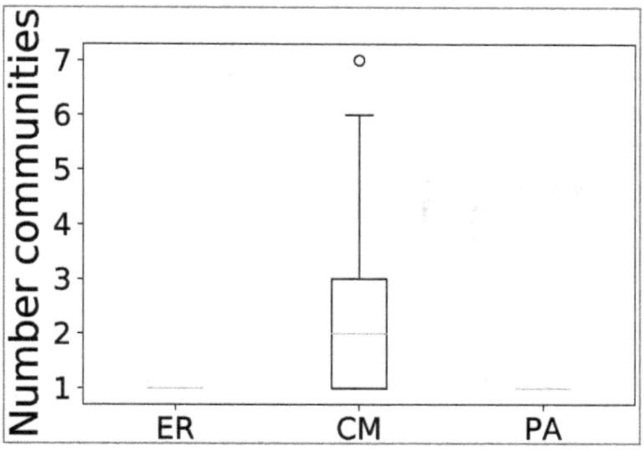

Figure 4.13. Boxplot of the number of clusters obtained by the Bayesian framework on different random graphs without community structure. The setting is the same as in Figure 4.10. We computed 100 random graphs with $n = 1500$ nodes. ER refers to Erdős-Rényi model with $p = \frac{4}{n}$, CM to a configuration model with a Zipf law of parameter 2 for degree distribution, and PA is the simple preferential attachment model as described in Section 2.2.2.

To finish this section, we show in Figure 4.13 that Bayesian clustering applied to random graph models with no community structure predicts in most situation only one community: the over-fitting issue is now absent.

4.4 Theoretical Analysis

4.4.1 Modularity and Maximum A Posteriori Estimator

In this section, we consider the adjacency matrix A of a random graph G sampled from a homogeneous degree-corrected block model, where the edges are Poisson distributed (see Section 2.3.2). More precisely, $A_{ii} = 0$ and for $i \neq j$

$$A_{ij} = A_{ji} \sim \begin{cases} \mathcal{P}(\theta_i\theta_j\omega_{\text{in}}), & \text{if } z_i^0 = z_j^0, \\ \mathcal{P}(\theta_i\theta_j\omega_{\text{out}}), & \text{otherwise,} \end{cases} \tag{4.19}$$

where $\mathcal{P}(\omega)$ denotes a Poisson random variable of parameter ω and d_i is the degree of node i. Similarly to the DC-SBM, we assume that for all $k \in [K]$, $\sum_i \theta_i 1\left(z_i^0 = k\right) = 1$. Proposition 4.4 shows that for such a block model, the *Maximum A Posteriori* (MAP) estimator defined by

$$\hat{z}^{\text{MAP}} = \arg\max_{z\in[K]^n} \mathbb{P}(z \mid A) \tag{4.20}$$

corresponds to maximising a quantity resembling the modularity.

Proposition 4.4. *Let A be the adjacency matrix of a block model graph with K blocks, n nodes, with uniform prior probability for the node labels and where the edges are* *sampled independently according to* (4.19). *Then, the MAP estimator defined in* (4.20) *verifies*

$$\hat{z}^{\mathrm{MAP}} = \underset{z\in[K]^n}{\arg\max} \sum_{i,j} \left(A_{ij} - \frac{\omega_{\mathrm{in}} - \omega_{\mathrm{out}}}{\log \frac{\omega_{\mathrm{in}}}{\omega_{\mathrm{out}}}} \theta_i \theta_j \right) 1\left(z_i = z_j\right).$$

Proof. The Bayes formula gives

$$\mathbb{P}\left(z \,|\, A\right) \propto \mathbb{P}\left(A \,|\, z\right) \mathbb{P}\left(z\right),$$

where the proportionality hides the term $\mathbb{P}(A)$ independent of z. Moreover,

$$\mathbb{P}\left(z\right) = \prod_{i=1}^{n} \mathbb{P}\left(z_i\right) = \frac{1}{K^n},$$

and hence $\mathbb{P}(z)$ is also independent of z. Therefore,

$$\underset{z\in[K]^n}{\arg\max}\, \mathbb{P}(z \,|\, A) = \underset{z\in[K]^n}{\arg\max}\, \mathbb{P}(A \,|\, z),$$

and $\mathbb{P}(A \,|\, z) = \prod_{i<j} \frac{\left(\theta_i\theta_j\omega_{ij}\right)^{A_{ij}}}{A_{ij}!} \mathrm{e}^{-\theta_i\theta_j\omega_{ij}}$, where

$$\omega_{ij} = \begin{cases} \omega_{\mathrm{in}}, & \text{if } z_i = z_j, \\ \omega_{\mathrm{out}}, & \text{otherwise.} \end{cases}$$

Thus,

$$\begin{aligned} \log \mathbb{P}(A \,|\, z) &= \sum_{i<j} \left(A_{ij} \log\left(\theta_i\theta_j\omega_{ij}\right) - \theta_i\theta_j\omega_{ij}\right) - \sum_{i<j} \log(A_{ij}!). \\ &= \frac{1}{2} \sum_{i\neq j} \left(A_{ij} \log\left(\theta_i\theta_j\omega_{ij}\right) - \theta_i\theta_j\omega_{ij}\right) - \sum_{i<j} \log\left(A_{ij}!\right). \end{aligned}$$

The last term $\sum_{i<j} \log(A_{ij}!)$ is independent of the model parameters and do not affect the position of the maximum. Moreover, we note that

$$\omega_{ij} = (\omega_{\mathrm{in}} - \omega_{\mathrm{out}}) 1\left(z_i = z_j\right) + \omega_{\mathrm{out}}.$$

(To show this, simply notice that the left hand side equals $(\omega_{\mathrm{in}} - \omega_{\mathrm{out}}) \times 0 + \omega_{\mathrm{out}} = \omega_{\mathrm{out}}$ when $z_i \neq z_j$, and equals $(\omega_{\mathrm{in}} - \omega_{\mathrm{out}}) \times 1 + \omega_{\mathrm{out}} = \omega_{\mathrm{in}}$ when $z_i \neq z_j$, hence it corresponds to the definition of ω_{ij}.) Similarly,

$$\begin{aligned} \log(\theta_i\theta_j\omega_{ij}) &= \left(\log(\theta_i\theta_j\omega_{\mathrm{in}}) - \log(\theta_i\theta_j\omega_{\mathrm{out}})\right) 1\left(z_i = z_j\right) + \log(\theta_i\theta_j\omega_{\mathrm{out}}) \\ &= \log \frac{\omega_{\mathrm{in}}}{\omega_{\mathrm{out}}} 1\left(z_i = z_j\right) + \log(\theta_i\theta_j\omega_{\mathrm{out}}). \end{aligned}$$

Therefore,

$$\log \mathbb{P}(A \mid z) = \frac{1}{2} \sum_{i \neq j} \left(A_{ij} \log \frac{\omega_{\text{in}}}{\omega_{\text{out}}} - (\omega_{\text{in}} - \omega_{\text{out}}) \theta_i \theta_j \right) 1 \left(z_i = z_j \right) + C,$$

where $C = \frac{1}{2} \sum_{i \neq j} \left(A_{ij} \log(\theta_i \theta_j \omega_{\text{out}}) - \theta_i \theta_j \omega_{\text{out}} \right) - \sum_{i<j} \log(A_{ij}!)$ is a constant term independent of z. Hence, we obtain

$$\log \mathbb{P}(A \mid z) = \frac{1}{2} \log \frac{\omega_{\text{in}}}{\omega_{\text{out}}} \sum_{i \neq j} \left(A_{ij} - \frac{\omega_{\text{in}} - \omega_{\text{out}}}{\log \frac{\omega_{\text{in}}}{\omega_{\text{out}}}} \theta_i \theta_j \right) 1 \left(z_i = z_j \right) + C$$

and this ends the proof. □

Recall that the modularity was defined by equation (4.17) as

$$\mathcal{M}(z) = \frac{1}{2|E|} \sum_{i,j} \left(A_{ij} - P_{ij} \right) 1(z_i = z_j),$$

where P_{ij} is the probability of an edge between i and j under a null-model, and the chosen null-model was the configuration model. Proposition 4.4 gives something similar. Indeed, we can write

$$\hat{z}^{\text{MAP}} = \arg\max_{z} \sum_{i,j} \left(A_{ij} - \gamma P_{ij} \right) 1(z_i = z_j),$$

where $\gamma = \frac{\omega_{\text{in}} - \omega_{\text{out}}}{\log \frac{\omega_{\text{in}}}{\omega_{\text{out}}}} \frac{K}{\omega_{\text{in}} + (K-1)\omega_{\text{out}}}$ and $P_{ij} = \theta_i \theta_j \frac{\omega_{\text{in}} + (K-1)\omega_{\text{out}}}{K}$ corresponds to the expected probability of observing an edge between i and j under the null-model (see Definition (4.19)). Moreover, the expected degree of a node i is equal to $\bar{d}_i = \sum_{j=1}^{n} \theta_i \theta_j \lambda_{ij} = \theta_i \frac{\omega_{\text{in}} + (K-1)\omega_{\text{out}}}{K}$, while the expected number of edges is equal to $\bar{m} = \frac{1}{2} \frac{\omega_{\text{in}} + (K-1)\omega_{\text{out}}}{K}$. Hence, $P_{ij} = \frac{\bar{d}_i \bar{d}_j}{2\bar{m}}$, and we recover

$$\hat{z}^{\text{MAP}} = \arg\max_{z \in [K]^n} \sum_{i,j} \left(A_{ij} - \gamma \frac{\bar{d}_i \bar{d}_j}{2\bar{m}} \right) 1 \left(z_i = z_j \right).$$

The quantity inside the arg max resembles to the modularity as defined in (4.17), with an extra parameter γ. One can define the *regularised modularity* (Reichardt and Bornholdt, 2006; Arenas *et al.*, 2008) as follows:

$$\mathcal{M}_\gamma(z) = \sum_{i,j} \left(A_{ij} - \gamma P_{ij} \right), \tag{4.21}$$

with P_{ij} usually taken equal to $\frac{d_i d_j}{2m}$.

Hence, the MAP estimator is equivalent to the maximisation of the generalized modularity, with $P_{ij} = \frac{\bar{d}_i \bar{d}_j}{2\bar{m}}$ and γ as defined earlier. Unfortunately, the relevance of this equivalence is rather limited, since the parameters $\omega_{\text{in}}, \omega_{\text{out}}$ (and hence γ) are usually unknown, albeit various strategies have been proposed for their estimation (see more on this in (Newman, 2016)).

4.4.2 Normalized Spectral Clustering as a Continuous Relaxation of Modularity Maximisation

The goal of this subsection is to link a particular relaxation of modularity maximisation to normalized spectral clustering. For simplicity of derivations, we will restrict the consideration to the case of two clusters. Recall that the maximisation of the generalised modularity is given by

$$\hat{z} = \underset{z \in \{-1,1\}^n}{\arg\max} \sum_{i,j} \left(A_{ij} - \gamma \frac{d_i d_j}{2|E|} \right) 1(z_i = z_j).$$

Noticing that $1(z_i = z_j) = \frac{1}{2}\left(z_i z_j + 1\right)$, we can rewrite it as

$$\hat{z} = \underset{z \in \{-1,1\}^n}{\arg\max} \sum_{i,j} B_{ij} z_i z_j,$$

where B is the matrix whose entries are $B_{ij} = A_{ij} - \gamma \frac{d_i d_j}{2|E|}$. As done for the cut-based methods (Section 4.1), we can simplify the problem by relaxing the discreteness of $z \in \{-1, 1\}^n$ to a real-valued vector $x \in \mathbb{R}^n$ (Newman, 2013). However, a constraint should be added to prevent x_i from becoming arbitrarily large, *i.e.*, to prevent the term $\left(A_{ij} - \gamma \frac{d_i d_j}{2|E|}\right) x_i x_j$ to become large in a trivial way. A straightforward constraint consists in fixing x onto the hyper-sphere by imposing $\sum_i x_i^2 = n$. This in particular sets the limit $-\sqrt{n} \leq x_i \leq \sqrt{n}$, while imposing the ℓ^2-norm of x to be equal to n. More generally, we can fix x to a hyper-ellipsoid by letting $\sum_i \kappa_i x_i^2 = \sum_i \kappa_i$ for a vector $\kappa = (\kappa_1, \ldots, \kappa_n)$ of non-negative entries. In particular, for $\kappa_i = d_i$ this leads to the following problem

$$\hat{x} = \underset{\substack{x \in \mathbb{R}^n \\ x^T D x = 2|E|}}{\arg\max} \; x^T B x,$$

where we used $x^T B x = \sum_{i,j} B_{ij} x_i x_j$.

The Lagrangian associated to the above problem is

$$x^T B x - \lambda \left(x^T D x - 2|E|\right),$$

and equating the derivative with respect to x to zero gives

$$Bx = \lambda Dx. \tag{4.22}$$

Thus x is a solution of a generalized eigenvector equation corresponding to an eigenvalue λ. To know which value of λ we should consider, we note that for a generalized eigenvector x, we have $x^T Bx = \lambda x^T Dx = \lambda 2|E|$, and hence the modularity $x^T Bx$ is highest for the largest eigenvalue λ of the generalized eigenproblem (4.22).

Since $B1_n = (1-\gamma)D1_n$, $\lambda = 1-\gamma$ is an admissible solution of (4.22). Therefore, if the maximum eigenvalue is $1-\gamma$, then the best partitioning corresponds to not dividing the network at all. We rule out this case, and thus assume $\lambda > 1-\gamma$. Noticing that $Bx = Ax - \gamma D1_n \frac{d^T x}{2|E|}$ with $d = (d_1, \ldots, d_n)$, we rewrite problem (4.22) as

$$Ax = D\left(\lambda x + \gamma 1_n \frac{d^T x}{2|E|}\right).$$

Multiplying on the left by 1^T leads to $d^T x = (\lambda + \gamma) d^T x = 0$ (we used $1^T A = 1^T D = d^T$ and $d^T 1 = 2|E|$). Since $\lambda > 1-\gamma$, this in turn implies $d^T x = 0$, and problem (4.22) simplifies to

$$Ax = \lambda Dx.$$

We notice that the constant vector $x = 1_n$ is a solution, and according to the Perron-Frobenius theorem it is associated to the largest eigenvalue since all its elements are positive. Nonetheless, we rule out this solution since it does not verify $d^T x = 0$, and we consider the second largest eigenvalue λ. Rescaling by $y = D^{1/2}x$ leads to the standard eigenvalue problem

$$D^{-1/2}AD^{-1/2}y = \lambda y,$$

or, equivalently,

$$\mathcal{L}y = (1-\lambda)y$$

to use the normalized Laplacian $\mathcal{L} = I_n - D^{-1/2}AD^{-1/2}$. Hence, y is an eigenvector of the normalized Laplacian. The link with normalized spectral clustering is completed by noticing that λ should be the second largest eigenvalue, and hence $1-\lambda$ should be the second smallest eigenvalue of $\mathcal{L}$.

4.4.3 Information-theoretic Results for Consistent Recovery in SBMs

This section presents information-theoretic results about recovery consistency in SBMs.

Non-binary SBMs

Let us first generalise SBMs to network with non-binary interactions. We denote by $\mathcal{S}$ the space of interactions, and by f_{in} and f_{out} the interaction densities (with respect to a measure μ). These parameters specify a probability measure on a space of observations

$$\mathcal{A} = \left\{ A = \left(a_{ij}\right) \in \mathcal{S}^{n \times n} \text{ such that } a_{ij} = a_{ji},\ a_{ii} = 0 \text{ for all } i, j \right\}$$

with probability density function

$$\mathbb{P}(A \,|\, z) = \prod_{1 \leq i < j \leq n} f_{z_i z_j}\left(a_{ij}\right) \tag{4.23}$$

with respect to the $n(n-1)/2$-fold product of the reference measure μ.

In other words, for an observation A distributed according to (4.23), the entries a_{ij}, $1 \leq i < j \leq n$, are mutually independent, and a_{ij} is distributed according to f_{in} when $z_i = z_j$, and according to f_{out} otherwise. In particular, when $\mathcal{S} = \{0, 1\}$ and $f_{\text{in}}, f_{\text{out}}$ are Bernoulli distributions, we recover the binary homogeneous SBM as defined in Section 2.3.1. When $\mathcal{S} = \mathbb{Z}$ and $f_{\text{in}}, f_{\text{out}}$ are Poisson distributions, we recover the Poisson SBM (see equation (2.7)).

The node labelling z representing the block membership structure is an unknown parameter to be estimated. We consider the node labelling as a random variable distributed according to the uniform distribution $\pi(z) = K^{-n}$ on the parameter space $\mathcal{Z} = \left\{ z \in [K]^n \right\}$. In this case the joint distribution of the node labelling and the observed data is characterised by a probability density

$$\mathbb{P}(\sigma, A) = \pi_z \mathbb{P}\left(A \,|\, z\right) \tag{4.24}$$

on $\mathcal{Z} \times \mathcal{X}$ with respect to $\text{card}_{\mathcal{Z}} \times \mu$, where $\text{card}_{\mathcal{Z}}$ is the counting measure on $\mathcal{Z}$.

Regime of asymptotic recovery

We recall that the *Hamming distance* between two sequences $y, z \in [K]^n$ is defined as the number of positions at which the corresponding symbols are different, *i.e.*,

$$d_{\text{Ham}}(y, z) = \sum_{i=1}^{n} 1(y_i \neq z_i).$$

For an estimator $\hat{z}$ of node labelling $z \in [K]^n$, we define the *absolute classification error* as follows:

$$d^*_{\mathrm{Ham}}\left(\hat{z}, z\right) = \min_{\tau \in \mathcal{S}_K} \sum_{i=1}^{n} 1\left(\tau \circ \left(\hat{z}_i\right) \neq z_i\right). \tag{4.25}$$

This corresponds to the number of misclassified nodes by the estimator $\hat{z}$ up to a global permutation[5] $\tau \in \mathcal{S}_K$.

When analysing the average performance of an estimator, we can view $\hat{z}\colon A \in \mathcal{A} \mapsto \hat{z}(A) \in [K]^n$ as a $[K]^n$-valued random variable defined on the set of observations $\mathcal{A}$. Then, $\mathbb{E}_z d^*_{\mathrm{Ham}}(\hat{z}, z)$ is equal to the expected clustering error given true node labeling z, and

$$\mathbb{E}d^*_{\mathrm{Ham}}(\hat{z}) = \sum_{z \in [K]^n} \pi(z) \mathbb{E}_z d^*_{\mathrm{Ham}}\left(\hat{z}, z\right)$$

is the average clustering error with respect to the node labelling distribution π on the parameter space.

We say that the estimator $\hat{z}$ asymptotically achieves *exact recovery*, or equivalently that $\hat{z}$ is a *strongly consistent estimator* of z, if

$$\mathbb{E}d^*_{\mathrm{Ham}}\left(\hat{z}\right) \to 0 \quad \text{as} \quad n \to \infty. \tag{4.26}$$

Condition (4.26) means that asymptotically every node is correctly classified. This demand is often excessive. A more reasonable setting is when only a vanishing fraction of nodes is misclassified (that is, at most $o(n)$ nodes are misclassified). We say that estimator $\hat{z}$ asymptotically achieves *almost exact recovery* (or that $\hat{z}$ is a *consistent estimator*) if

$$n^{-1}\mathbb{E}d^*_{\mathrm{Ham}}\left(\hat{z}\right) \to 0 \quad \text{as} \quad n \to \infty.$$

Remark 4.4. Exact and almost exact recovery are two the most studied regimes of cluster recovery. Another regime is called detection and is much weaker (it corresponds to the regime when there exists an estimator performing better than a random guess). This condition is weaker and therefore holds even if the graph is very sparse (for example, when the average degree is constant). We will not discuss this regime here, since the proof techniques are very different. We refer the reader to Moore, 2017.

---

5. The presence of the global permutation in this definition is justified by the observation that permuting all the node labels do not change the overall clustering.

Information-theoretic conditions for consistent recovery

Definition 4.3 (Rényi divergence)**.** The *Rényi divergence* between two probability distributions f and g is defined as

$$D_{1/2}(f,g) = -2\log \int \left(\frac{df}{d\mu}\right)^{1/2} \left(\frac{dg}{d\mu}\right)^{1/2} d\mu,$$

where μ is an arbitrary measure which dominates f and g. We use the following conventions: $\log 0 = -\infty$, $0/0 = 0$, and $x/0 = \infty$ for $x > 0$.

Remark 4.5. The Rényi divergence is linked to the *Hellinger distance*, $\mathrm{Hel}(f,g)$, defined by $\mathrm{Hel}^2(f,g) = \frac{1}{2}\int \left(\sqrt{\frac{df}{d\mu}} - \sqrt{\frac{dg}{d\mu}}\right)^2 d\mu$, via the formula $D_{1/2}(f,g) = -2\log\left(1 - \mathrm{Hel}^2(f,g)\right)$.

In what follows, we assume that a sigma-finite reference measure μ on $\mathcal{S}$ is fixed once and for all, and we write $\frac{df}{d\mu}, \frac{dg}{d\mu}$ simply as f, g, and we omit $d\mu$ from the integral signs, so that $D_{1/2}(f,g) = -2\log\int\sqrt{fg}$. When $\mathcal{S}$ is countable, μ is always chosen as the counting measure, in which case we write $D_{1/2}(f,g) = -2\log\sum_{x\in\mathcal{S}}\sqrt{f(x)g(x)}$.

Theorem 4.6. *Consider a homogeneous SBM with $n \gg 1$ nodes, $K \asymp 1$ blocks, and interaction distributions $f_{\mathrm{in}} = f_{\mathrm{in}}^{(n)}$ and $f_{\mathrm{out}} = f_{\mathrm{out}}^{(n)}$ over $\mathcal{S} = \mathcal{S}^{(n)}$. Let $I = I_n$ be the Rényi divergence between f and g. The following holds:*

(i) *a consistent estimator exists if $I \gg n^{-1}$ and does not exist if $I \lesssim n^{-1}$;*
(ii) *a strongly consistent estimator exists if $I \geq (1+\Omega(1))\frac{K\log n}{n}$, and does not exist if $I \leq (1-\Omega(1))\frac{K\log n}{n}$.*

Theorem 4.6 shows that the Rényi divergence governs the possibility or impossibility of (strongly) consistent recovery in non-binary SBMs. In fact, when interaction distributions f_{in} and f_{out} are too similar (in the sense that their Rényi divergence is smaller than n^{-1}), then there is not enough information provided by the network to recover the communities consistently.

Theorem 4.6 is proved in Avrachenkov *et al.*, 2022. The literature on consistency thresholds in SBMs is large, and one can refer to Zhang *et al.*, 2016 for binary SBMs ($\mathcal{S} = \{0,1\}$) and to Jog and Loh, 2015; Xu *et al.*, 2020 for weighted ($\mathcal{S} = \mathbb{R}$) or edge-labelled ($\mathcal{S} = \{0,1,\cdots,L\}$) SBMs.

Application to binary SBMs

Let us see how we can apply Theorem 4.6 to sparse binary SBMs, for which $f_{\text{in}} = \text{Ber}(p_{\text{in}})$ and $f_{\text{out}} = \text{Ber}(p_{\text{out}})$ with $p_{\text{in}}, p_{\text{out}} \ll 1$. A Taylor expansion gives

$$
\begin{aligned}
D_{1/2}(f_{\text{in}}, f_{\text{out}}) &= -2\log\left(\sqrt{(1-p_{\text{in}})(1-p_{\text{out}})} + \sqrt{p_{\text{in}}p_{\text{out}}}\right) \\
&= -2\log\left(1 - \frac{p_{\text{in}} + p_{\text{out}}}{2} + \sqrt{p_{\text{in}}p_{\text{out}}} + O(p_{\text{in}}p_{\text{out}})\right) \\
&= -2\log\left(1 - \frac{\left(\sqrt{p_{\text{in}}} - \sqrt{p_{\text{out}}}\right)^2}{2} + O(p_{\text{in}}p_{\text{out}})\right) \\
&= \left(\sqrt{p_{\text{in}}} - \sqrt{p_{\text{out}}}\right)^2 + O(p_{\text{in}}p_{\text{out}}). \qquad (4.27)
\end{aligned}
$$

This can be applied to the following two particular cases.

Example 4.1. In a regime when $p_{\text{in}} = a\rho_n$ and $p_{\text{out}} = b\rho_n$ for scale-independent constants $a \neq b$ and with $\rho_n \ll 1$. Theorem 4.6 and equation (4.27) tell that a consistent estimator exists if $n\rho_n \gg 1$, and does not exist if $n\rho_n \lesssim 1$. We note that the key quantity $n\rho_n$ is of the same order as the expected degree $\bar{d}_n = \frac{a+b}{2}n\rho_n$. Thus, for the possibility of consistent recovery we require that the expected degree diverges with the size of the network.

Example 4.2. In a regime where $p_{\text{in}} = a\frac{\log N}{N}$ and $p_{\text{out}} = b\frac{\log N}{N}$ for scale-independent constants a, b, Theorem 4.6 and equation (4.27) tell that a strongly consistent estimator exists if $(\sqrt{a}-\sqrt{b})^2 > K$ and does not exist if $(\sqrt{a}-\sqrt{b})^2 < K$. This is the well-known threshold for strong consistency in binary SBMs (Abbe *et al.*, 2015; Mossel *et al.*, 2015).

Remark 4.6. Considering the setting of Example 4.2, we see that for $K = 2$ strong consistency requires $\frac{\left(\sqrt{a}-\sqrt{b}\right)^2}{2} = \frac{a+b}{2} - \sqrt{ab} > 1$. Since $\frac{a+b}{2} > 1$ is the condition for connectivity in this SBM (see Theorem 2.2), this means that exact recovery in SBM is a strictly stronger requirement than connectivity.[6]

Other Particular Cases of Non-binary SBMs

Example 4.3 (Poisson interactions)**.** The Rényi divergence between Poisson distributions with means λ and μ is exactly equal to $I = (\sqrt{\lambda} - \sqrt{\mu})^2$. In a regime when $\lambda = a\frac{\log n}{n}$ and $\mu = b\frac{\log n}{n}$ with constants $a, b > 0$, Theorem 4.6 tells that

6. If the graph was not connected, then a.s. the graph would contain isolated nodes (see Lemma 2.4). Hence, exact recovery would not be possible, as one could not correctly classify the isolated nodes better than a random guess. Therefore exact recovery requires connectivity, but Example 4.2 shows that connectivity alone is not enough.

a strongly consistent estimator exists if $(\sqrt{a}-\sqrt{b})^2 > K$ and does not exist if $(\sqrt{a}-\sqrt{b})^2 < K$. This condition is similar to the condition in Example 4.2, and is due to the fact that Poisson distributions with small mean are well approximated by Bernoulli distributions.

Example 4.4 (Censored block model)**.** Let us consider a latent binary SBM with $f_{\text{in}} = \text{Ber}(p_0)$ and $f_{\text{out}} = \text{Ber}(q_0)$ for which each interaction and non-interaction are revealed independently with probability $r = r_0 \frac{\log n}{n}$, where we assume that p_0, q_0 and r_0 are constants. The resulting observed network is a non-binary SBM with interaction space $\mathcal{S} = \{\text{present}, \text{absent}, \text{censored}\}$ (where censored denotes the unobserved interactions) and with intra-block and inter-block probability distributions $\tilde{f}_{\text{out}}$ and $\tilde{f}_{\text{in}}$. We have $\tilde{f}_{\text{out}}(\text{present}) = rp_0, \tilde{f}_{\text{out}}(\text{absent}) = r(1-p_0)$ and $\tilde{f}_{\text{out}}(\text{censored}) = 1-r$, and similarly for $\tilde{f}_{\text{in}}$. From $D_{1/2}(\tilde{f}_{\text{out}}, \tilde{f}_{\text{in}}) = r\left(\left(\sqrt{p_0}-\sqrt{q_0}\right)^2 + \left(\sqrt{1-p_0}-\sqrt{1-q_0}\right)^2\right) + O\left(r^2\right)$ it follows that a strongly consistent estimator exists if $r_0 > r_0^{\text{crit}}$ and does not exist if $r_0 < r_0^{\text{crit}}$, where $r_0^{\text{crit}} = \frac{K}{\left(\sqrt{p_0}-\sqrt{q_0}\right)^2+\left(\sqrt{1-p_0}-\sqrt{1-q_0}\right)^2}$. For $K = 2$, this coincides with the critical threshold obtained in Dhara *et al.*, 2022.

4.4.4 Consistency of Spectral Methods in SBM

In this section, we will prove that spectral clustering is consistent in the SBM. For simplicity, we will consider spectral clustering using the graph adjacency matrix, but a similar proof would hold if one were to use the normalized Laplacian.

Heuristic: mean-field model

We first consider the mean-field model of the SBM, that is the model where all the random quantities are replaced by their expectations. In particular, the mean-field graph becomes the weighted graph formed by the expected adjacency matrix of a SBM graph. Therefore, if (z, G) is drawn from $\text{SBM}(n, \pi, Q)$, then the adjacency matrix of the corresponding mean-field is

$$\mathbb{E}A = ZQZ^T,$$

where $Q \in [0,1]^{K \times K}$ is the rate matrix (recall that element $Q_{k\ell}$ denotes the probability of edge appearance between a node in community k and a node in community ℓ), and $Z \in \{0,1\}^{n \times K}$ is the *membership matrix* defined by

$$Z_{ik} = \begin{cases} 1, & \text{if} \quad z_i = k, \\ 0, & \text{otherwise.} \end{cases}$$

The following lemma specifies the eigenstructure of $\mathbb{E}A$.

Lemma 4.7. *Assume Q is full-rank, and let UDU^T be an eigendecomposition of $\mathbb{E}A$. Then $U = ZX$, where $X \in \mathbb{R}^{K \times K}$ and $\|X_{k*} - X_{\ell*}\| = \sqrt{n_k^{-1} + n_\ell^{-1}}$ for all $1 \leq k < \ell \leq K$, and where X_{k*} denotes the row k of X.*

Proof. Let $\Delta = \text{diag}(\sqrt{n_1}, \ldots, \sqrt{n_K})$. Then, we can write

$$\mathbb{E}A = ZQZ^T = (Z\Delta^{-1})(\Delta Q \Delta)(Z\Delta^{-1})^T.$$

The matrix $Z\Delta^{-1}$ is orthonormal. Indeed,

$$(Z\Delta^{-1})^T Z\Delta^{-1} = \Delta^{-1}Z^TZ\Delta^{-1} = I_n,$$

where we used the fact that $Z^TZ = \text{diag}(n_1, \ldots, n_K) = \Delta^2$.

Let RDR^T be the eigendecomposition of $\Delta Q \Delta$. Thus,

$$\mathbb{E}A = (Z\Delta^{-1}R)\, D\, (Z\Delta^{-1}R)^T$$

is the eigendecomposition of $\mathbb{E}A$. We finish the proof by letting $U = Z\Delta^{-1}R$ and $X = \Delta^{-1}R$. We then have

$$XX^T = \text{diag}\left(n_1^{-1}, \ldots, n_K^{-1}\right).$$

Hence,

$$\begin{aligned}\|X_{k*} - X_{\ell*}\| &= \|X_{k*}\| + \|X_{\ell*}\| - 2X_{k*}X_{\ell*}^T \\ &= n_k^{-1} + n_\ell^{-1} + 0,\end{aligned}$$

and the claim follows. □

In particular, Lemma 4.7 ensures that the community information is encoded in the eigenstructure of $\mathbb{E}A$. Indeed, the K eigenvectors associated with non-zero eigenvalues of $\mathbb{E}A$ are given by the columns of U, which can be written as ZX. The k-means step (see equation (4.10)) then aims to recover Z (and X) from U.

Consistency of spectral clustering in SBM

We established that if one were to observe the mean-field graph, then recovery of communities would be possible by looking at the K leading eigenvectors of the mean-field adjacency matrix $\mathbb{E}A$. The following theorem states that, under some natural conditions, consistent recovery is possible by looking at the K leading eigenvectors of the random graph adjacency matrix A. We recall that the absolute classification error $d^*_{\text{Ham}}(\hat{z}, z)$ is defined in (4.25), and an estimator is consistent if $\frac{d^*_{\text{Ham}}(\hat{z},z)}{n} = o(1)$.

Theorem 4.8. *Let* $(z, G) \sim \text{SBM}(n, \pi, P)$*, where* P *is of rank* K *with smallest absolute nonzero eigenvalue larger than* γ_n*. Let* $\bar{d}_n$ *be the expected degree and* $\hat{z} \in [K]^n$ *be the output of spectral clustering applied to the adjacency matrix. Then, there exists a constant* $c > 0$ *such that if* $(2+\epsilon)\frac{K\bar{d}_n}{\gamma_n^2} < c$*, then with high probability*

$$\frac{d^*_{\text{Ham}}(\hat{z}, z)}{n} \leq (2+\epsilon)^2 c \frac{K\bar{d}_n}{\gamma_n^2}.$$

Example 4.5. Consider a homogeneous SBM with $P_{k\ell} = p_{\text{in}}$ if $k = \ell$ and $P_{k\ell} = p_{\text{out}}$ otherwise. Then, $\bar{d}_n = \frac{n}{K}\left(p_{\text{in}} + (K-1)p_{\text{out}}\right)$ while $\gamma_n = \frac{n}{K}\left(p_{\text{in}} - p_{\text{out}}\right)$. Assume that $p_{\text{in}} = c_{\text{in}}\rho_n$, $p_{\text{out}} = c_{\text{out}}\rho_n$ where $c_{\text{in}}, c_{\text{out}}$ does not depend on n, and suppose that the assumptions of Theorem 4.8 hold. Then, the error of spectral clustering is bounded by

$$\frac{d^*_{\text{Ham}}(\hat{z}, z)}{n} \leq (2+\epsilon)^2 cK \left(\frac{c_{\text{in}} + (K-1)c_{\text{out}}}{c_{\text{in}} - c_{\text{out}}}\right)^2 \frac{1}{\bar{d}_n}.$$

This upper bound goes to zero when the average degree $\bar{d}_n$ goes to infinity, ensuring consistency of spectral methods in such a setting.

The intuition for the proof of Theorem 4.8 is as follows.

- Show that the K leading eigenvectors of the adjacency matrix A are not too different from the K leading eigenvectors of the expected adjacency matrix $\mathbb{E}A$. This is done in two steps.
  - First use a result from random matrix theory to show that A is concentrated around $\mathbb{E}A$. This is Theorem 4.9.
  - Then use this concentration to show that the eigenvectors are also concentrated. This is usually done using Davis-Kahan theorem. We present a variation of it in Lemma 4.10.
- Conclude by bounding the error made by the k-means step.

Theorem 4.9 (Theorem 1.2 of Le *et al.*, 2017). *Let* A *be the adjacency matrix of a Bernoulli random graph* $\mathcal{G}(n, (p_{ij}))$*, and let* $d_n = n \max_{ij} p_{ij}$*. For* $\tau \sim d$*, define* $A_\tau = A + \tau 1_n 1_n^T$ *the regularized adjacency matrix. Then, we have with high probability when* n *goes to infinity*

$$\|A_\tau - \mathbb{E}A_\tau\|_2 = O\left(\sqrt{d_n}\right).$$

The proof of Theorem 4.9 is complex and out of reach for this book. We will simply note that the regularization term $\tau 1_n 1_n^T$ is needed to ensure concentration of the adjacency matrix when d_n is small. Indeed, consider an Erdős-Rényi graph by letting $p_{ij} = p$. If $d_n \ll \log n$, then the degree of some nodes are much larger than the expected degree $d_n = np$. This implies that some rows of the adjacency matrix will have ℓ^2 norms much larger than d_n, which in turn imply $\|A - \mathbb{E}A\| \gg \sqrt{d_n}$.

Lemma 4.10 (Principal subspace perturbation). *Let $\bar{M} \in \mathbb{R}^{n\times n}$ be a symmetric matrix with smallest nonzero singular value γ, and let M be any symmetric matrix. Denote by U and $\bar{U} \in \mathbb{R}^{n\times K}$ the matrices whose columns are composed of the K leading eigenvectors of M and $\bar{M}$. Then, there exists a $K \times K$ orthogonal matrix Q such that*

$$\left\| \bar{U}Q - U \right\|_F \leq \frac{2\sqrt{2K}}{\gamma} \left\| M - \bar{M} \right\|_2 .$$

Lemma 4.10 is a version of Davis-Kahan 'sin θ' theorem that bounds the distance between two subspaces spanned by the leading eigenvectors of two matrices. We refer to the Theorem 2 of Yu *et al.*, 2015 for further explanations.

The last ingredient needed for the proof of Theorem 4.8 is a bound on the error made by the k-means step. The next lemma gives such a bound.

Lemma 4.11 (Approximate k-means error bound, adapted from Lemma 5.3 of Lei and Rinaldo, 2015). *For $\epsilon > 0$ and any matrices $\bar{V}, V \in \mathbb{R}^{n\times K}$ such that $\bar{V} = ZX$ with $Z \in \mathcal{Z}_{n,K}$, and $X \in \mathbb{R}^{K\times K}$, let $\left(\widehat{Z}, \widehat{X}\right)$ be a $(1+\epsilon)$ approximation of the k-means problem (4.10). We denote by z and $\hat{z}$ the membership vectors associated to the membership matrices Z and $\widehat{Z}$. Let $n_{\min}$ be the size of the smallest community, and $\delta = \min_{k,\ell:\, k\neq\ell} \|X_{k*} - X_{\ell*}\|$. If $4(2+\epsilon)\frac{\|V-\bar{V}\|_F^2}{\delta^2} \leq n_{\min}$, then*

$$\frac{d^*_{\mathrm{Ham}}(\hat{z}, z)}{n} \leq 4(2+\epsilon)^2 \frac{\|V - \bar{V}\|_F^2}{\delta^2 n}.$$

Lemma 4.11 upper bounds the error made by the k-means step. The bound involves the Frobenius distance between $\bar{V}$ (the matrix with the eigenvectors of $\mathbb{E}A$, which we know from the mean-field study Lemma 4.7, can be written as ZX and from which we can recover the community structure Z), and the matrix V (the matrix with the eigenvectors of A).

Proof. Denote by $\widehat{V} = \widehat{Z}\widehat{X}$. Intuitively, we want to show that if V is close to $\bar{V}$, then $\widehat{V}$ is close to $\bar{V}$ as well, where $\widehat{V}$ is the solution of the minimisation problem (4.10) with the objective function V. Let $\mathcal{C}_k := \{i\colon z_i = k\}$ be the set of nodes belonging to community k, and $\mathcal{B}_k := \{i \in \mathcal{C}_k\colon \|\bar{V}_{i*} - (\widehat{Z}\widehat{X})_{i*}\|_2 \geq \delta/2\}$. The sets $\mathcal{B}_k$ corresponds to nodes for which the k-means solution $\widehat{V} = \widehat{Z}\widehat{X}$ is far away from the matrix $\bar{V}$. Let us first show that the sets $\mathcal{B}_k$ are of small sizes. We have

$$\begin{aligned}\|\bar{V} - \widehat{Z}\widehat{X}\|_F^2 &= \sum_{i=1}^{n}\left(\sum_{j=1}^{n}\left|\bar{V} - (\widehat{Z}\widehat{X})_{ij}\right|^2\right)\\ &= \sum_{i}\|\bar{V}_{i*} - (\widehat{Z}\widehat{X})_{i*}\|^2\\ &= \sum_{k=1}^{K}\sum_{i\in\mathcal{C}_k}\|\bar{V}_{i*} - (\widehat{Z}\widehat{X})_{i*}\|^2,\end{aligned}$$

and thus,

$$\|\bar{V} - \widehat{Z}\widehat{X}\|_F^2 \geq \sum_{k=1}^{K}\sum_{i\in\mathcal{B}_k}\|\bar{V}_{i*} - (\widehat{Z}\widehat{X})_{i*}\|^2 \geq \frac{\delta^2}{4}\sum_{k=1}^{K}|\mathcal{B}_k|.$$

Hence,

$$\begin{aligned}\sum_{k=1}^{K}|\mathcal{B}_k| &\leq \frac{4}{\delta^2}\|\bar{V} - \widehat{Z}\widehat{X}\|_F^2\\ &\leq \frac{4}{\delta^2}\left(\|\bar{V} - V\|_F + \|V - \widehat{Z}\widehat{X}\|_F\right)^2\\ &\leq \frac{4}{\delta^2}\left(1 + \sqrt{1+\epsilon}\right)^2\|V - \bar{V}\|_F^2\\ &\leq \frac{4}{\delta^2}(2+\epsilon)^2\|V - \bar{V}\|_F^2,\end{aligned}\tag{4.28}$$

since $\|\widehat{Z}\widehat{X} - V\|_F^2 \leq (1+\epsilon)\|Z'X' - V\|^2$ for all $Z', X' \in \mathcal{Z}_{n,K} \times \mathbb{R}^{K\times K}$.

Using the assumption of the lemma, we have $\sum_{k=1}^{K}|\mathcal{B}_k| < n_{\min}$. Hence, for every $k \in [K]$, the sets $\mathcal{C}_k\backslash\mathcal{B}_k$ are non-empty. We will now claim that:

(i) If $i \in \mathcal{C}_k\backslash\mathcal{B}_k$ and $j \in \mathcal{C}_\ell\backslash\mathcal{B}_\ell$ with $k \neq \ell$, then $\widehat{V}_{i*} \neq \widehat{V}_{j*}$;
(ii) For $i, j \in \mathcal{C}_k\backslash\mathcal{B}_k$, we have $\widehat{V}_{i*} = \widehat{V}_{j*}$.

Therefore, every node $i \notin \cup_{k=1}^{K}\mathcal{B}_k$ can be assigned to a class $\hat{z}_i$ based on the value of the row i of $\widehat{V}$. Let $\sigma^* \in \mathcal{S}_K$ be a permutation satisfying

$$\sigma^* \in \arg\min_{\sigma\in\mathcal{S}_K}\sum_{i\notin\cup_{k=1}^{K}\mathcal{B}_k} 1(\sigma(\hat{z}_i) \neq z_i).$$

For such σ^*, we have $\sum_{i=1}^{n} 1(\sigma(\hat{z}_i) \neq z_i) \le \sum_{k=1}^{K} |\mathcal{B}_k|$. Thus,

$$d^*_{\mathrm{Ham}}(\hat{z}, z) \le \sum_{i=1}^{n} 1(\sigma(\hat{z}_i) \neq z_i) \le \sum_{k=1}^{K} |\mathcal{B}_k|,$$

and we conclude that the lemma's claim is true using equation (4.28).

Let us now show claim (*i*). If we were to assume that $\widehat{V}_{i*} = \widehat{V}_{j*}$, then this would imply $\delta \le \|\bar{V}_{i*} - \bar{V}_{j*}\|_2 \le \|\bar{V}_{i*} - \hat{V}_{i*}\|_2 + \|\widehat{V}_{j*} - \bar{V}_{j*}\|_2 < \delta/2 + \delta/2$, which is a contradiction.

Finally, let us show claim (*ii*). Since $\widehat{Z} \in \mathcal{Z}_{n,K}$ and $\widehat{X} \in \mathbb{R}^{K\times K}$, then $\widehat{V}$ has at most K distinct rows. We also know from claim (*i*) that $\widehat{V}$ has at least K distinct rows. Hence, $\widehat{V}$ has exactly K distinct rows, and this in turn implies that $\widehat{V}_{i*} = \widehat{V}_{j*}$ for $i, j \in \mathcal{C}_k \backslash \mathcal{B}_k$. □

Proof of Theorem 4.8. Let V (resp. $\bar{V}$) be a n-by-K matrix whose columns are composed of the K leading eigenvectors of A_τ (resp. of $\mathbb{E}A_\tau$). Combining Lemma 4.10 and Theorem 4.9, we have for some orthogonal matrix $Q \in \mathbb{R}^{K\times K}$

$$\|\bar{V}Q - V\|_F \le \frac{2\sqrt{2K}}{\gamma_n}\|A_\tau - \mathbb{E}A_\tau\|_2 \le \frac{2\sqrt{2K}}{\gamma_n}C\sqrt{\bar{d}}, \tag{4.29}$$

with high probability.

We will now directly apply Lemma 4.11 to V and $\bar{V}Q$. Lemma 4.7 shows that $\bar{V}Q = ZXQ = ZX'$ with $X' = XQ$, where $\|X'_{k*} - X_{\ell*}\|_2^2 = \sqrt{\frac{1}{n_k} + \frac{1}{n_\ell}}$. Therefore, we can choose $\delta = 1/\sqrt{n_{\max}}$. Using equation (4.29), a sufficient condition for $4(2+\epsilon)\frac{\|V-\bar{V}\|_F^2}{\delta^2} \le n_{\min}$ to hold is

$$4(2+\epsilon)8C^2K\frac{\bar{d}}{\gamma_n^2} \le n_{\min}n_{\max}.$$

Therefore, we can apply Lemma 4.11, which states that

$$\frac{d^*_{\mathrm{Ham}}(\hat{z}, z)}{n} \le 4(2+\epsilon)^2\frac{\|V - \bar{V}Q\|_F^2}{\delta^2 n} \le 4(2+\epsilon)^2 8C^2K\frac{\bar{d}_n}{\delta^2 n\gamma_n^2},$$

and the statement of the theorem holds, since $\frac{1}{\delta^2 n} \le 1$. □

Further Notes

Spectral clustering is well explained in the review by Von Luxburg, 2007. For more details on Louvain algorithm, we refer the reader to Blondel *et al.*, 2008; Good *et al.*, 2010. A nice application of Louvain algorithm (and more generally community detection methods) to content recommendations on Reddit is presented in Jamonnak *et al.*, 2015. Additional deficiencies of Louvain algorithm (such as badly connected communities) were discovered by Traag *et al.*, 2019, who also proposed a refinement of Louvain algorithm (called *Leiden algorithm*). We also mention the *resolution limit* problem (Fortunato and Barthelemy, 2007), common to modularity maximisation methods. We finally note that while modularity methods are popular, thanks to the existence of fast algorithms and to the heuristic consideration linking modularity maximisation with maximum likelihood approach, care is needed since modularity maximisation is not strictly equivalent to likelihood maximisation (Zhang and Peixoto, 2020) and modularity algorithms are prone to over-fitting.

Consistency of spectral methods in SBM were studied by Lei and Rinaldo, 2015, and further developed in Abbe *et al.*, 2020. For a recent overview of various applications of spectral methods, we refer to Chen *et al.*, 2021. Spectral methods are not the only ones to be consistent on SBM. For example, the consistency of SDP methods has been demonstrated (Hajek *et al.*, 2016a,b; Guédon and Vershynin, 2016; Amini *et al.*, 2018; Fei and Chen, 2019). Moreover, community detection in other block models such as the Geometric Block Model has been recently studied (Galhotra *et al.*, 2018; Sankararaman and Baccelli, 2018; Avrachenkov *et al.*, 2021a).

Many other community detection methods exist, for example: belief propagation (Moore, 2017; Decelle *et al.*, 2011), game-theoretic methods (Avrachenkov *et al.*, 2018a; Moscato *et al.*, 2019), methods based on the map equation (Rosvall and Bergstrom, 2008; Rosvall *et al.*, 2009) and spectral methods based on other matrices such as the non-backtracking matrix (Krzakala *et al.*, 2013) or the Bethe-Hessian (Saade *et al.*, 2014). We also refer to the review by Fortunato, 2010 for more insights about the community detection problem.

Finally, an important question not covered here is the estimation of the number of communities. For this topic, we refer the reader to Le and Levina, 2015; Bickel and Sarkar, 2016; Lei, 2016; Saldana *et al.*, 2017; Hu *et al.*, 2020.

DOI: 10.1561/9781638280514.ch5

Chapter 5

Graph-based Semi-supervised Learning

Semi-supervised learning (SSL) aims at achieving superior learning performance by combining unlabelled and labelled data. Since typically the amount of unlabelled data is large compared to the amount of labelled data, SSL methods are relevant when the performance of unsupervised learning is low, or when the cost of getting a large amount of labelled data for supervised learning is too high. Unfortunately, many standard semi-supervised learning techniques have been shown to not efficiently use the unlabelled data, leading to unsatisfactory or unstable performances (Chapelle *et al.*, 2006, Chapter 4; Ben-David *et al.*, 2008; Cozman *et al.*, 2002). Moreover, the presence of noise in the labelled data may further degrade their performance. In practice, the noise often comes from a tired or non-diligent expert carrying out the labelling task.

In this chapter, we will review some standard methods for semi-supervised graph clustering. In particular, we will study the performance of those methods in the case when the amount of labelled data is low and we will propose robust solutions in the presence of noisy labels.

General idea We assume that the node set $V = [n]$ of a graph $G = (V, E)$ is partitioned into K non-overlapping communities, represented by the latent community labelling vector $z \in [K]^n$. It will be convenient to have a *one-hot* representation of z, by defining a $n \times K$ ground-truth *membership matrix* $Z \in \{0,1\}^{n \times K}$, such that

$$Z_{ik} = \begin{cases} 1, & \text{if } z_i = k, \\ 0, & \text{otherwise.} \end{cases}$$

As seen in Chapter 4, unsupervised community detection is the problem of recovering Z from the observation of G (and sometimes with the knowledge of K). We study here the noisy semi-supervised setting. More precisely, we assume that, in addition to the observation of the graph, an *oracle* gives us extra information about the cluster assignment of some nodes. We call those nodes the *labelled nodes*, and we denote by ℓ the set of labelled nodes. Among those nodes, some are correctly labelled by the oracle, while some are mislabelled by the oracle. We denote by ℓ_0 the set of mislabelled nodes and ℓ_1 the set of correctly labelled nodes. In particular, $\ell = \ell_0 \sqcup \ell_1$. The oracle can be represented by a matrix S of size $n \times K$, whose rows $S_{i.}$ are given by

$$S_{i.} = \begin{cases} Z_{i.}, & \text{if } \quad i \in \ell_1, \\ \widetilde{Z}_{i.}, & \text{if } \quad i \in \ell_0, \\ 0_{1\times K}, & \text{if } \quad i \notin \ell, \end{cases} \tag{5.1}$$

where $\widetilde{Z}_{i.}$ is chosen in $\{z \in \{0,1\}^K : \|z\|_1 = 1 \text{ and } z \neq Z_{i.}\}$, and $0_{1\times K}$ denotes the row of K zeros.

In other words, the oracle (5.1) reveals the correct cluster assignment of $|\ell_1|$ nodes, and a false cluster assignment for $|\ell_0|$ nodes. It reveals nothing for $n - |\ell|$ nodes. The quantity $|\ell_0|/|\ell|$ is the rate of mistakes of the oracle (*i.e.*, the probability that the oracle reveals a false information given that it reveals something). The oracle is informative if this quantity is less than $1/2$, which is equivalent to the intuitive condition $|\ell_1| > |\ell_0|$. In the following, we will always assume that the oracle is informative.

Assumption 5.1. The oracle is informative, that is $|\ell_1| > |\ell_0|$.

Given the oracle S and the graph G, our strategy is to find a matrix $\widehat{X} \in \mathbb{R}^{n\times K}$ from which we could predict the nodes' labels. We will refer to the rows $X_{.k}$ as classification functions, and a node i will be classified in cluster $\hat{z}_i$ if

$$\hat{z}_i = \underset{k\in\{1,\dots,K\}}{\arg\max}\, X_{ik}. \tag{5.2}$$

A standard framework is to define $\widehat{X}$ as the solution of an optimisation problem of the type

$$\widehat{X} = \underset{X \in \mathcal{X}}{\arg\min}\, C(X, S),$$

where $C(X, S)$ is a cost function, and $\mathcal{X}$ is a subset of $\mathbb{R}^{n \times K}$.

Notations Throughout this chapter, ℓ denotes the set of nodes labelled by the oracle, while $u = [n] \backslash \ell$ denotes the set of unlabelled nodes. The oracle is represented by a matrix $S \in \{0, 1\}^{n \times K}$, defined as in equation (5.1), and the goal is to infer $Z \in \{0, 1\}^{n \times K}$ after the observation of the graph G and the oracle S.

We can assume, up to a reordering of the nodes, that the first $|\ell|$ nodes are labelled by the oracle, while the remaining $|u|$ ones are not. Accordingly, any matrix $M \in \mathbb{R}^{n \times n}$ can be displayed in the block form

$$M = \begin{pmatrix} M_{\ell\ell} & M_{\ell u} \\ M_{u\ell} & M_{uu} \end{pmatrix}.$$

Moreover, for any matrix $X = (X_{ik}) \in \mathbb{R}^{n \times K}$, $X_{i\cdot}$ stands for the row i of X, while $X_{\cdot k}$ stands for the column k of X, and we write $X = \begin{pmatrix} X_{\ell\cdot} \\ X_{u\cdot} \end{pmatrix}$.

Finally, I_ℓ denotes the diagonal matrix whose element (i, i) equals 1 if $i \in \ell$ and 0 otherwise.

5.1 Laplacian-based SSL Methods

5.1.1 Label Propagation

Presentation of the method Spectral methods for unsupervised community detection are based on the minimisation of quadratic functions, such as $\operatorname{Tr}\left(X^T L X\right)$ or $\operatorname{Tr}\left(X^T \mathcal{L} X\right)$ (see Section 4.1). *Label Propagation* extends this to the semi-supervised setting. In particular, the foundational papers (Zhu and Ghahramani; Zhu *et al.*, 2003) considered the following optimisation problem

$$\widehat{X}^{LP} = \underset{\substack{X \in \mathbb{R}^{n \times K} \\ X_{\ell\cdot} = S_{\ell\cdot}}}{\arg\min} \operatorname{Tr}\left(X^T L X\right). \tag{5.3}$$

The constraint $X_{\ell\cdot} = S_{\ell\cdot}$ forces the solution $\widehat{X}^{LP}$ to be equal to the oracle prediction on the labelled nodes. We first note that this *hard constraint* may not be suitable if the oracle is noisy, as it pushes the solution on the wrongly labelled nodes towards a wrong classification. Moreover, the constraints $X^T X = I_K$ or $X^T D X = I_K$

(see again Section 4.1) of spectral clustering, which prevent obtaining flat solutions in unsupervised spectral methods, are absent here. Hence, the optimisation problem (5.3) relies only on the hard constraint $X_{\ell\cdot} = S_{\ell\cdot}$ to prevent degenerate solutions. We will see later that this is a problem when the amount of labelled data is small. On the positive side, the following lemma provides a closed-form expression for $\widehat{X}^{LP}$.

Lemma 5.2 The solution $\widehat{X}^{LP}$ of the optimisation problem (5.3) is given by

$$\begin{cases} \widehat{X}^{LP}_{\ell\cdot} &= S_{\ell\cdot}, \\ \widehat{X}^{LP}_{u\cdot} &= \left(I_{|u|} - \left(D^{-1}A\right)_{uu}\right)^{-1} \left(D^{-1}A\right)_{u\ell} S_{\ell\cdot}, \end{cases} \tag{5.4}$$

where $I_{|u|}$ is the identity matrix of size $|u| \times |u|$.

Proof. The constraint $X_{\ell\cdot} = S_{\ell\cdot}$ can be rewritten as follows:

$$\begin{aligned} X_{\ell\cdot} = S_{\ell\cdot} &\iff \forall k \in [K]\ \forall i \in \ell:\ (X_{ik} - S_{ik})^2 = 0 \\ &\iff \sum_{k=1}^{K}\sum_{i=1}^{n} (1(i \in \ell)X_{ik} - S_{ik})^2 = 0 \\ &\iff \operatorname{Tr}\left((I_\ell X - S)^T (I_\ell X - S)\right) = 0. \end{aligned}$$

Thus, a Lagrangian associated to the minimisation problem (5.3) is

$$\mathcal{H} = \operatorname{Tr}\left(X^T L X + \mu (I_\ell X - S)^T (I_\ell X - S)\right),$$

where μ is a Lagrange multiplier. For every $k \in [K]$, the derivative with respect to $X_{\cdot k}$ gives

$$\frac{\partial \mathcal{H}}{\partial X_{\cdot k}} = 2\left(LX + \mu (I_\ell X - S)\right).$$

Equating this derivative to zero leads to

$$(L + \mu I_\ell)\widehat{X}^{LP} = \mu S,$$

while the derivative with respect to μ leads to the constraint $I_\ell \widehat{X}^{LP} = S$. Using block notation, we can write

$$LX = \begin{pmatrix} L_{\ell\ell} & L_{\ell u} \\ L_{u\ell} & L_{uu} \end{pmatrix} \begin{pmatrix} X_{\ell\cdot} \\ X_{u\cdot} \end{pmatrix} = \begin{pmatrix} L_{\ell\ell}X_{\ell\cdot} + L_{\ell u}X_{u\cdot} \\ L_{u\ell}X_{\ell\cdot} + L_{uu}X_{u\cdot} \end{pmatrix},$$

and therefore

$$\begin{cases} L_{\ell\ell}\widehat{X}_{\ell\cdot}^{LP} + L_{\ell u}\widehat{X}_{u\cdot}^{LP} + \mu\widehat{X}_{\ell\cdot}^{LP} & = \mu S_{\ell\cdot}, \\ L_{u\ell}\widehat{X}_{\ell\cdot}^{LP} + L_{uu}\widehat{X}_{u\cdot}^{LP} & = 0. \end{cases}$$

The constraint $\widehat{X}_{\ell\cdot}^{LP} = S_{\ell\cdot}$ leads to the solution

$$\begin{cases} \widehat{X}_{\ell\cdot}^{LP} & = S_{\ell\cdot}, \\ \widehat{X}_{u\cdot}^{LP} & = (L_{uu})^{-1} L_{u\ell}\widehat{X}_{\ell\cdot}^{LP}. \end{cases}$$

We end the proof by noticing that since $L = D - A$ and D is a diagonal matrix, we have $L_{u\ell} = -A_{u\ell}$ and $(L_{uu})^{-1} = \left((D(I_n - D^{-1}A))_{uu}\right)^{-1} = \left(I_{|u|} - (D^{-1}A)_{uu}\right)^{-1}(D_{uu})^{-1}$. Finally, $(D_{uu})^{-1}A_{u\ell} = (D^{-1}A)_{u\ell}$, since D is diagonal, and this ends the proof. □

We note from the proof of Lemma 5.2 that

$$LX_{ik} = \begin{cases} S_{ik}, & \text{if } i \in \ell, \\ 0, & \text{otherwise.} \end{cases} \tag{5.5}$$

Finally, we present the following Algorithm 8. The computation of $\widehat{X}$ from equation (5.4) requires to solve a $|u|$-by-$|u|$ linear system, which has in general a time-complexity of $O\left(|u|^3\right)$ (less if the network is sparse). The ensuing paragraph presents a method to compute $\widehat{X}$ in a decentralized and iterative manner.

Algorithm 8: Label Propagation (Zhu and Ghahramani; Zhu *et al.*, 2003).

Input: graph G, oracle S.
Output: node labelling $\hat{z} = (\hat{z}_1, \ldots, \hat{z}_n) \in [K]^n$.
Process:

- let $\widehat{X}$ as in equation (5.4);
- for $i = 1, \ldots, n$ let $\hat{z}_i$ be defined by classification rule (5.2).

Return: $\hat{z}$.

Interpretation as a propagation of the oracle labels We start by assigning to each node i a 1-by-K vector $X_{i\cdot}^{(0)} \in \mathbb{R}^{1\times K}$, which is equal to the oracle prediction $S_{i\cdot}$ for node i. Then, at each time step t, the update of X is done as follows:

- if $i \in \ell$, then $X_{i\cdot}^{(t+1)} = X_{i\cdot}^{(t)}$ (no update is done);

- if $i \notin \ell$, then $X_{i\cdot}^{(t+1)}$ is the taken to be the average of $X_{j\cdot}^{(t)}$ over the neighbours of node i, that is, $X_{i\cdot} = \frac{1}{d_i}\sum_{j=1}^{n} A_{ij}X_{j\cdot}^{(t)}$.

This can be interpreted as a propagation of the oracle's information through the graph or as a consensus algorithm with the states of some agents fixed. The labelled nodes' value remains equal to the oracle information, while an unlabelled node will sample the value of its neighbours and perform local averaging. In matrix form, we can write this procedure as follows:

$$X_{i\cdot}^{(t+1)} = \begin{cases} X_{i\cdot}^{(t)}, & \text{if } i \in \ell, \\ \left(D^{-1}AX^{(t)}\right)_{i\cdot}, & \text{otherwise.} \end{cases}$$

The use of block notation leads to

$$\begin{cases} X_{u\cdot}^{(t+1)} &= \left(D^{-1}A\right)_{uu} X_{u\cdot}^{(t)} + \left(D^{-1}A\right)_{u\ell} X_{\ell\cdot}^{(t)}, \\ X_{\ell\cdot}^{(t+1)} &= X_{\ell\cdot}^{(t)}, \end{cases}$$

with the initial condition $X^{(1)} = S$. Since the matrix $\left(D^{-1}A\right)_{uu}$ is substochastic, $X^{(t)}$ converges to X^{∞} satisfying the following system of equations

$$\begin{cases} X_{u\cdot}^{\infty} &= \left(D^{-1}A\right)_{uu} X_{u\cdot}^{\infty} + \left(D^{-1}A\right)_{u\ell} X_{\ell\cdot}^{\infty}, \\ X_{\ell\cdot}^{\infty} &= S_{\ell\cdot}, \end{cases}$$

whose solutions are

$$\begin{cases} X_{u\cdot}^{\infty} &= \left(I_{|u|} - \left(D^{-1}A\right)_{uu}\right)^{-1} \left(D^{-1}A\right)_{u\ell} S_{\ell\cdot}, \\ X_{\ell\cdot}^{\infty} &= S_{\ell\cdot}, \end{cases}$$

which is the same expression as the Label Propagation solution (5.4).

Random walk interpretation Let $y_1, y_2, \dots$ be a random walk on the graph, where the walker jumps from node i to a node j, where j is a neighbour of i chosen uniformly at random. The transition probabilities are given by

$$p_{ij} = \mathbb{P}\left(y_{t+1} = j \,|\, y_t = i\right) = \frac{A_{ij}}{d_i},$$

where d_i is the degree of node i. In particular, $p_{ij} = \left(D^{-1}A\right)_{ij}$ and we note that $P = D^{-1}A$ is the matrix of transition probabilities.

Suppose that the walk starts from node i, and that we end the walk as soon as we hit a labelled node. We denote y_{end} the final node. We denote by $\widehat{X}_{ik}$ the probability that $S_{y_{\text{end}},k} = 1$, that is the oracle assigned y_{end} to community k by the oracle. Thus, we have

$$\widehat{X}_{ik} = \mathbb{P}\left(S_{y_{\text{end}},k} = 1 \mid y_1 = i\right).$$

In particular, if $i \in \ell$, then $y_{\text{end}} = i$ and

$$\widehat{X}_{ik} = \begin{cases} 1, & \text{if } S_{ik} = 1, \\ 0, & \text{otherwise,} \end{cases}$$

which is equivalent to $\widehat{X}_{\ell\cdot} = S_{\ell\cdot}$. By Markov property, we also have for any node i

$$\mathbb{P}\left(S_{y_{\text{end}},k} = 1 \mid y_1 = i\right) = \sum_{j=1}^{n} \mathbb{P}\left(S_{y_{\text{end}},k} = 1 \mid y_1 = j\right) p_{ij},$$

and therefore

$$\widehat{X} = P\widehat{X}.$$

Writing this equation in the block form and combining it with the constraint $\widehat{X}_{\ell\cdot} = S_{\ell\cdot}$ obtained before, leads to

$$\widehat{X}_{u\cdot} = \left(I_{|u|} - \left(D^{-1}A\right)_{uu}\right)^{-1} \left(D^{-1}A\right)_{u\ell} S_{\ell\cdot}.$$

Hence, we again recover the same expression as the solution of Label Propagation (5.4).

Interpretation as a heat equation Let us now interpret the Label Propagation as a solution of a *heat equation*. The evolution of temperature T of an isotropic material is governed by the heat equation

$$\frac{\partial T}{\partial t} = \alpha \Delta T,$$

where Δ is the Laplacian and α is the thermal conductivity of the material. At equilibrium, we simply have $\Delta T = 0$.

The oracle S plays the role of a heat bath. More precisely, we first fix a $k \in [K]$. The labelled nodes $i \in \ell$ behave as heat sources, whose temperature T_{ik} remains constant and equal to $S_{ik} \in \{0, 1\}$. The temperatures of the unlabelled nodes vary, as heat exchange takes place along the graph edges and is proportional to the temperature difference between the edges' endpoints. Therefore,

$$\forall i \in \ell : \; T_{ik} = S_{ik},$$
$$\forall i \in u : \; \frac{\partial T_{ik}}{\partial t} = \sum_{j=1}^{n} A_{ij}(T_{jk} - T_{ik}).$$

Since $\sum_{j=1}^{n} A_{ij}(T_{jk} - T_{ik}) = (AT_{\cdot k})_i - d_i T_{ik} = -(LT_{\cdot k})_i$, the temperature $T_{\cdot k}$ verifies at equilibrium

$$\forall i \in u : \; LT_{ik} = 0,$$

while $T_{ik} = S_{ik}$ for any labelled node i. This can be rewritten as

$$\begin{cases} (LT)_{u\cdot} = 0, \\ T_{\ell\cdot} = S_{\ell\cdot}. \end{cases}$$

The above system is equivalent to equation (5.5), whose solution is equal to the solution of Label Propagation (5.4) (see Lemma 5.2).

5.1.2 Label Spreading

The SSL method of *Label Spreading* (Zhou *et al.*, 2004) is based on the optimisation problem

$$\widehat{X}^{LS} = \underset{X \in \mathbb{R}^{n \times K}}{\arg\min}\, C^{LS}(X),$$

where the cost function C^{LS} is defined by

$$C^{LS}(X) = \mathrm{Tr}\left(X^T \mathcal{L} X + \lambda (X - S)^T (X - S)\right).$$

After simple linear algebra transformation, we have

$$C^{LS}(X) = \sum_{k=1}^{K} \left(\frac{1}{2} \sum_{i,j} a_{ij} \left(\frac{x_{ik}}{\sqrt{d_i}} - \frac{x_{jk}}{\sqrt{d_j}} \right)^2 + \lambda \sum_{i=1}^{n} (x_{ik} - s_{ik})^2 \right),$$

where d_i denotes the degree of node i.

The parameter λ enforces a trade-off between the smoothness of the solution $\widehat{X}_{LS}$ over the graph and the closeness of the solution to the oracle information S. The difference with the Label Propagation method is that the smoothness of the solution is imposed by the term $\mathrm{Tr}(X^T \mathcal{L} X)$, which now includes the normalization by the node degrees.

As in the case of Label Propagation, $\widehat{X}^{LS}$ can also be expressed in a closed form. Namely, for each $k \in [K]$, we have

$$\frac{1}{2}\frac{\partial C^{LS}}{\partial X_{\cdot k}} = \mathcal{L}X_{\cdot k} + \lambda\,(X_{\cdot k} - S_{\cdot k}),$$

and hence

$$\begin{aligned}\widehat{X}_{\cdot k}^{LS} &= (\lambda I + \mathcal{L})\,\lambda S_{\cdot k} \\ &= \left((1+\lambda)I - D^{-1/2}AD^{-1/2}\right)^{-1}\lambda S_{\cdot k} \\ &= \frac{\lambda}{1+\lambda}\left(I - \frac{1}{1+\lambda}D^{-1/2}AD^{-1/2}\right)^{-1} S_{\cdot k}.\end{aligned}$$

Therefore,

$$\widehat{X}^{LS} = (1-\alpha)\left(I - \alpha D^{-1/2}AD^{-1/2}\right)^{-1} S,$$

where $\alpha = \frac{\lambda}{1+\lambda} \in (0,1)$. This gives Algorithm 9.

Algorithm 9: Label Spreading (Zhou *et al.*, 2004).

Input: graph G, oracle S, parameter $\alpha \in (0,1)$.
Output: node labelling $\hat{z} = (\hat{z}_1, \ldots, \hat{z}_n) \in [K]^n$.
Process:

- compute the normalized adjacency matrix $\mathcal{A} = D^{-1/2}AD^{-1/2}$;
- let $\widehat{X}^{LS}$ be the solution of $(I - \alpha\mathcal{A})\,\widehat{X}^{LS} = (1-\alpha)S$;
- for $i \in [n]$, let $\hat{z}_i$ be defined by classification rule (5.2).

Return: $\hat{z}$

5.1.3 Generalized Laplacian

As a follow-up to Label Propagation and Label Spreading methods, Avrachenkov *et al.*, 2012 proposed a general class of cost functions

$$C^{GL}(X) = \mathrm{Tr}\left(X^T D^{\sigma-1} L D^{\sigma-1} X + \lambda\,(X-S)^T D^{2\sigma-1}(X-S)\right),$$

where $\lambda > 0$ and $0 \le \sigma \le 1$ are two hyper-parameters. The solution of the minimisation problem

$$\widehat{X}^{GL} = \operatorname*{arg\,min}_{X \in \mathbb{R}^{n\times K}} C^{GL}(X)$$

is given by

$$\widehat{X}^{GL} = (1-\alpha)\left(I_n - \alpha D^{-\sigma} A D^{\sigma-1}\right)^{-1} S,$$

where $\alpha = \lambda/(1+\lambda)$. Since the computations are similar to the computations in the previous sections, we omit them and refer the reader to (Avrachenkov *et al.*, 2012, Proposition 2) for details. Different normalizations are obtained by different choices of σ. In particular,

- $\sigma = 1$ corresponds to Label Propagation;
- $\sigma = 1/2$ corresponds to Label Spreading;
- $\sigma = 0$ corresponds to the PageRank-based method.

5.1.4 Numerical Performance of the Laplacian-based Methods

Choice of hyper-parameter α for Label Spreading We first investigate the effect of α on the classification performances. We choose two datasets for which we saw that unsupervised spectral clustering failed: *DBLP* and *Cora*. We let 2% of the nodes be labelled by the oracle, and we plot in Figure 5.1 (blue curve) the accuracy as a function of α. We see that the accuracy increased when α increases, but suddenly dropped if α becomes too close to one. We also notice that the best choice of α can be made by looking at the modularity of the predicted partition (red curve in Figure 5.1), as the modularity closely follows the accuracy.

Noisy oracle We now study the effect of noise on classification performance. We keep the same two datasets, now with 5% labelled nodes. We define the noise as being the proportion of mistakes made by the oracle. The results are plotted in Figure 5.2. As expected, the noise deteriorates the classification performance.

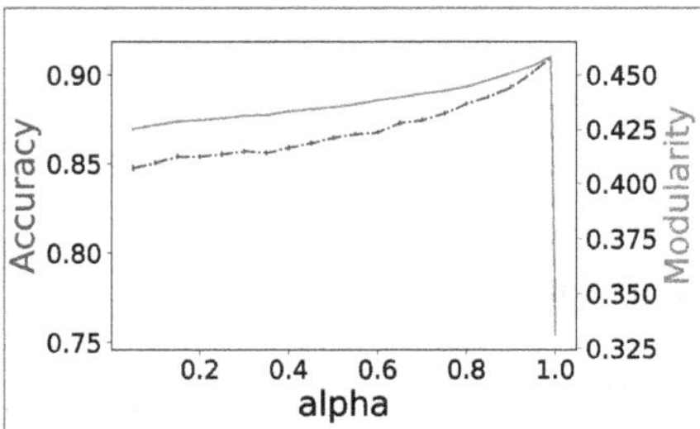

(a) *DBLP* dataset.

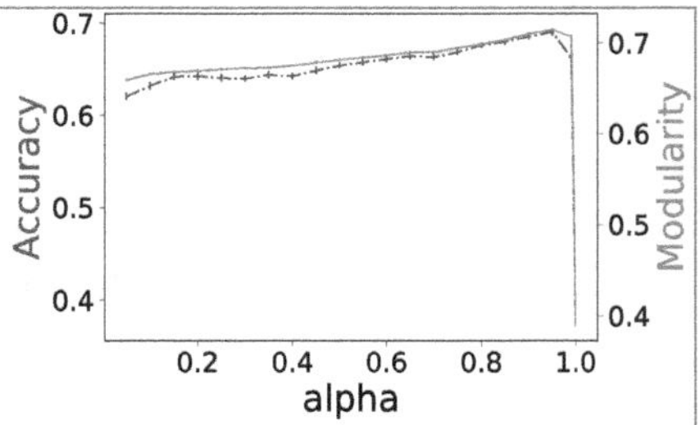

(b) *Cora* dataset.

Figure 5.1. Effect of the choice of parameter α on the performance of Label Spreading on two data sets. The blue curve gives the accuracy (computed with respect to the ground truth labels) and the red curve the modularity (computed using only the observed graph and the predicted labels). Results are averaged over 100 realisations. In each realisation, we randomly chose 2% of the nodes to serve as labelled nodes.

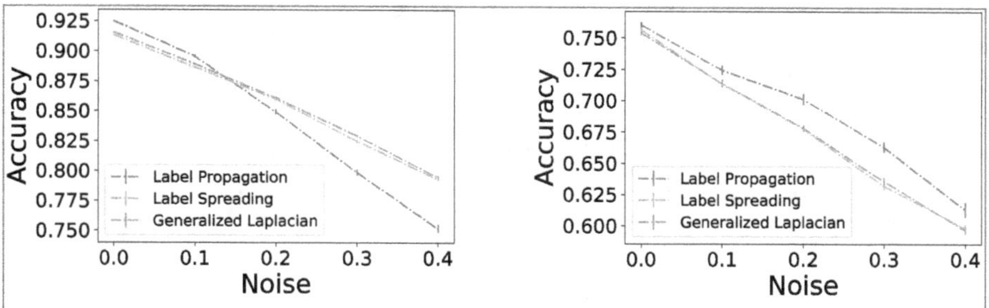

(a) *DBLP* dataset. (b) *Cora* dataset.

Figure 5.2. Effect of a noisy oracle on the classification performances of Laplacian-based methods. Results are averaged over 100 realisations, with 5% of the nodes being labelled. (We choose $\alpha = 0.8$ for Label Spreading and Generalized Laplacian, and $\sigma = 0$ for Generalized Laplacian, which corresponds to the PageRank-based method.)

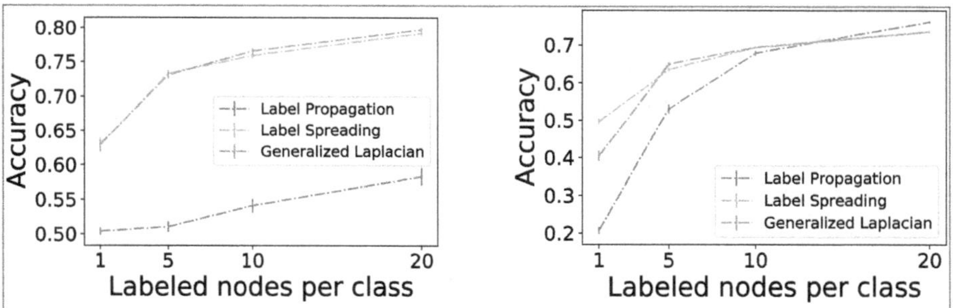

(a) *DBLP* dataset. (b) *Cora* dataset.

Figure 5.3. Effect of small labelled data on the classification performance of Laplacian-based methods. Results are averaged over 100 realisations.

Small amount of labelled data We finish this section by emphasizing the importance of the problem of small amount of labelled data. Figure 5.3 shows that the classification accuracy heavily degrades when the number of labelled nodes per class becomes too low.

5.2 Learning with Small Amount of Labelled Data

5.2.1 The Problem of Small Labelled Data

Numerical experiments showed that at very low labelling rates, the performance of the SSL methods becomes poor. We will explain this phenomenon using the random walk interpretation of Label Propagation algorithm (see Section 5.1.1 for more details on Label Propagation).

Let $y_1, \ldots, y_t, \ldots$ be a random walk on the graph starting at node i. Let $\tau = \inf_{t \geq 1}\{y_t \in \ell\}$ be the first time that the walk hits a labelled node, and we recall that

$\widehat{X}_{ik}^{LP} = \mathbb{P}\left(S_{y_\tau,k} = 1 \mid y_1 = i\right)$. In other words, $\widehat{X}_{ik}$ is the probability that the first labelled node reached by the walk (started from node i) has a label k.

If the number of labelled nodes is small and the graph is large, then the time τ will be large. In particular, if τ is larger than the *mixing time* of the walk, then the distribution of y_τ is very close to the invariant distribution π of the random walk, given by

$$\pi_j = \frac{d_j}{\sum_{s=1}^{n} d_s}.$$

This means that the chain has forgotten its starting point i, and thus $\widehat{X}_{ik}^{LP}$ is a constant independent of i, which defeats the goal of classification.

Let us formalize this intuition and try to mitigate the problem. We first note that, for $k \leq \tau$,

$$\mathbb{E}\left[X_{y_k} - X_{y_{k-1}} \mid y_{k-1}\right] = \frac{1}{d(y_{k-1})} LX_{y_{k-1}} = 0,$$

since $LX = 0$ on the unlabelled nodes (see equation (5.5)). Thus, $X_{y_1}, \cdots, X_{y_k}, \cdots$ is a martingale. Since τ is an almost surely bounded stopping time, Doob's optimal stopping theorem then implies that

$$\mathbb{E}\left[X_{y_0}\right] = \mathbb{E}\left[X_{y_\tau}\right].$$

Since $y_0 = i$ and $y_\tau \in \ell$, we have $\mathbb{E}\left[X_{y_0}\right] = X_i$ and $X_{y_\tau} = S_{y_\tau}$, and thus

$$X_{ik} \approx \sum_{j \in \ell} \pi_j S_{jk} = \frac{\sum_{j \in \ell} d_j S_{jk}}{\sum_{j=1}^{n} d_j}. \tag{5.6}$$

Hence, the first order approximation of $\widehat{X}_{ik}^{LP}$ is the same for all unlabelled node i, and potential differences only come from second-order terms. A first improvement of Label Propagation is thus to replace the classification rule (5.2) by

$$\hat{z}_i = \underset{k \in \{1,\ldots,k\}}{\arg\max} \left(\widehat{X}_{ik} - c_k\right),$$

where $c_k = \frac{\sum_{j \in \ell} d_j S_{jk}}{\sum_{j \in \ell} d_j}$. Equivalently, one could "shift" equation (5.5) and solve

$$LX_{ik} = \begin{cases} S_{ik} - c_k, & \text{if } i \in \ell, \\ 0, & \text{otherwise.} \end{cases}$$

Rewriting the above equation as

$$LX_{ik} = \sum_{j\in\ell} d_j \left(S_{jk} - c_k\right) \delta_{ij},$$

we can interpret it as a heat equation, where heat sources and sinks are placed at the labelled nodes.

5.2.2 Poisson Learning

Let $\bar{s}_k = \frac{\sum_{i\in\ell} S_{ik}}{|\ell|}$. Following the previous remarks, Calder *et al.*, 2020 proposed to consider the equation

$$LX_{ik} = \sum_{j\in\ell} \left(S_{jk} - \bar{s}_k\right) \delta_{ij}, \tag{5.7}$$

such that $\sum_{i=1}^{n} d_i X_{ik} = 0$. Equivalently, this accounts to solve the following optimisation problem (Calder *et al.*, 2020, Theorem 2.3)

$$\underset{\substack{X\in\mathbb{R}^{n\times K}\\ \sum_{i=1}^{n} d_i X_{ik}=0}}{\arg\min} \quad \mathrm{Tr}\left(X^T L X\right) - \left(S - \bar{S}\right)^T X,$$

where

$$\bar{S}_{ik}, = \begin{cases} \bar{s}_k & \text{if} i \in \ell, \\ 0, & \text{otherwise.} \end{cases}$$

In particular, while Label Propagation handles labelled data by placing hard constraints, Poisson learning adds a loss term to the energy function.

Random walk interpretation As the labelled nodes are now source and sinks of the heat equation, the random walk interpretation differs. Let us denote by $y_1^j, \ldots, y_t^j, \ldots$ a random walk on the graph starting from node $j \in \ell$. Each time the random walk hits node i, we record the shifted label $S_{j\cdot} - \bar{S}_{j\cdot}$. This defines the quantity

$$X_{ik}^{(T)} = \mathbb{E}\left[\sum_{t=0}^{T} \frac{1}{d_i} \sum_{j\in\ell} \left(S_{jk} - \bar{S}_{jk}\right) 1\left(y_t^j = i\right)\right].$$

The following lemma gives an iterative expression for $X^{(T)}$.

Lemma 5.3. *We have*

$$X_{ik}^{(T+1)} = X_{ik}^{(T)} + \frac{1}{d_i}\left(\sum_{j\in\ell}\left(S_{jk} - \bar{S}_{jk}\right)\delta_{ij} - \left(LX^{(T)}\right)_{ik}\right).$$

Furthermore, assume G is connected and the Markov chain induced by the random walk is aperiodic. Then $\lim_{T\to\infty} X^{(T)} = X$, where X is the unique solution of the Poisson equation (5.7).

Proof. We first write

$$X_{ik}^{(T+1)} = \sum_{j\in\ell}\left(S_{jk} - \bar{S}_{jk}\right) G_T(i,j), \tag{5.8}$$

where $G_T(i,j) = \frac{1}{d_i}\mathbb{E}\left[\sum_{t=0}^{T} 1\left(y_t^j = i\right)\right] = \frac{1}{d_i}\sum_{t=0}^{T}\mathbb{P}\left(y_t^j = i\right)$ is the normalized Green function. Using

$$\mathbb{P}\left(y_t^j = i\right) = \sum_{u=1}^{n}\mathbb{P}\left(y_t^j = i \,|\, y_{t-1}^j = u\right)\mathbb{P}\left(y_{t-1}^j = u\right),$$

we have

$$\begin{aligned} d_i G_T(i,j) &= \delta_{ij} + \sum_{t=1}^{T}\sum_{u=1}^{n}\frac{w_{ui}}{d_u}\mathbb{P}\left(y_{t-1}^j = u\right) \\ &= \delta_{ij}\sum_{u=1}^{n}\frac{w_{ui}}{d_u}\sum_{t=0}^{T-1}\mathbb{P}\left(y_t^j = u\right) \\ &= \delta_{ij} + \sum_{u=1}^{n} w_{ui} G_{T-1}(u,j), \end{aligned}$$

and therefore

$$d_i\left(G_T(i,j) - G_{T-1}(i,j)\right) + LG_{T-1}(i,j) = \delta_{ij}.$$

Combined with equation (5.8), this establishes

$$d_i\left(X_{ik}^{(T)} - X_{ik}^{(T-1)}\right) = \sum_{j\in\ell}\left(S_{jk} - \bar{S}_{jk}\right)\delta_{ij} - \left(LX^{(T)}\right)_{ik}.$$

Then, summing both sides of this equation over $i = 1, \cdots, n$, leads to

$$\sum_{i=1}^{n} d_i X_{ik}^{(T)} = \sum_{i=1}^{n} d_i X_{ik}^{(T-1)},$$

and therefore $\sum_{i=1}^{n} d_i X_{ik}^{(T)} = \sum_{i=1}^{n} d_i X_{ik}^{(0)}$ for all T. Since

$$d_i X_{ik}^{(0)} = \sum_{j \in \ell} \left(S_{jk} - \bar{S}_{jk} \right) \delta_{ij},$$

we obtain $\sum_i d_i X_{ik}^{(T)} = 0$. Finally, let $V_{ik}^{(T)} = d_i (X_{ik}^{(T)} - X_{ik})$. We have

$$V_{ik}^{(T)} = \sum_{j=1}^{n} \frac{w_{ij}}{d_j} V_{jk}^{(T-1)},$$

and $\sum_{j=1}^{n} V_{jk}^{(T)} = 0$ for all T. Since the random walk is aperiodic and the graph is connected,

$$\lim_{T \to \infty} V_{ik}^{(T)} = \pi_i \sum_{j=1}^{n} V_{jk}^{(0)} = 0,$$

where $\pi_i = \frac{d_i}{\sum_{j=1}^{n} d_j}$ is the chain's stationary distribution. □

In fact, Lemma 5.3 also provides an iterative numerical procedure for computing the solution. The precedure is formally described in Algorithm 10.

Algorithm 10: Poisson learning (Calder *et al.*, 2020).

Input: graph G, oracle $S \in \{0,1\}^{n \times K}$, number of iterations T.
Output: node labelling $\hat{z} \in [K]^n$.
Process:

- let L be the graph's standard Laplacian and D the graph's degree matrix;
- let ℓ be the set of labelled nodes, and let $\bar{S} = S \operatorname{diag}(\bar{s})$ where $\bar{s} = (\bar{s}_1, \ldots, \bar{s}_K)$ with $\bar{s}_k = \frac{1}{|\ell|} \sum_{i \in \ell} S_{i\ell}$;
- for $t = 1, \cdots, T$ do: $X \leftarrow X + D^{-1} \left(S - \bar{S} - LX \right)$;
- for $i = 1, \ldots, n$ let $\hat{z}_i$ be defined by classification rule (5.2).

Return: $\hat{z}$

5.2.3 Numerical Experiments

To assess the performance of Poisson learning in a regime with extremely low amount of labelled nodes, we reproduce the results of Calder *et al.*, 2020. They consider MNIST (LeCun *et al.*, 1998) and Fashion-MNIST (Xiao *et al.*, 2017) datasets, on which they trained auto-encoders to extract important features from

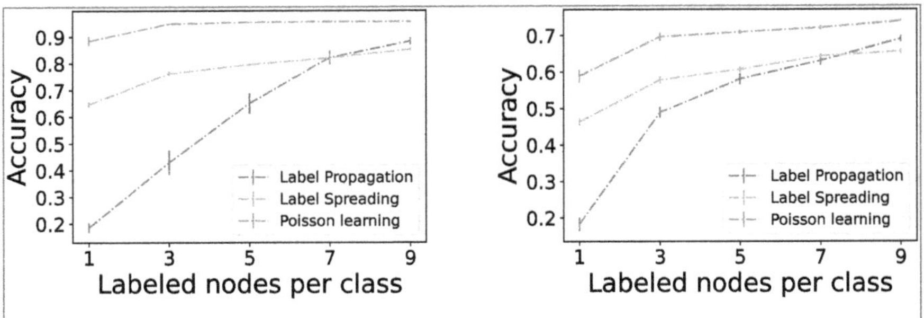

(a) *MNIST* dataset. (b) *Fashion-MNIST* dataset.

Figure 5.4. Performance of Poisson learning on MNIST and fashion-MNIST datasets, when the number of labelled nodes per class is extremely small. Results are averaged over 10 realisations, and error bars show the standard error.

the data. More precisely, they used variational auto-encoders with 3 fully connected layers of sizes (784,400,20) and (784,400,30), respectively, followed by a symmetrically defined decoder (Kingma and Welling, 2014). The auto-encoder was trained for 100 epochs on each data set. Then, a 10-nearest neighbours graph is build with Gaussian weights $w_{ij} = \exp\left(-4\|x_i - x_j\|^2/\sigma_i^2\right)$, where x_i is the latent variables for image i and σ_i is the distance between x_i and its 10-nearest neighbour. The results are shown in Figure 5.4. In particular, even with only one labelled node per class, the performance of *Poisson learning* remains extremely high.

5.3 Other Methods

5.3.1 Constrained Spectral Clustering

The goal of this method is to directly incorporate semi-supervised information into spectral methods. Specifically, from the oracle information S, we can construct the must-link/cannot-link matrix Q as follows:

$$Q_{ij} = Q_{ji} = \begin{cases} 1, & \text{if } i,j \in \ell \text{ and } S_{i\cdot} = S_{j\cdot}, \\ -1, & \text{if } i,j \in \ell \text{ and } S_{i\cdot} \neq S_{j\cdot}, \\ 0, & \text{otherwise.} \end{cases}$$

We note that must-link/cannot-link information can also be given to us directly. In fact, for experts it is often easier to conclude if two items are similar or not, rather than to attribute items to classes.

For a membership matrix $Z \in \mathcal{Z}_{N,K}$, the quantity

$$\mathrm{Tr}\left(Z^T QZ\right) = \sum_{k=1}^{K}\sum_{i,j} Q_{ij}Z_{ik}Z_{jk}$$

measures how well the membership matrix Z respects the oracle information. Indeed, this quantity increases by 1 if $Q_{ij} = 1$ and Z assign nodes i, j in the same cluster, and decreases by 1 if $Q_{ij} = -1$ but i and j are assigned to the same cluster. Therefore, $\mathrm{Tr}\left(Z^T QZ\right)$ equals the number of satisfied must-link/cannot-link constraints minus the number of violated constraints. Rather than asking for all the constraints in Q to be verified, we can impose the lower bound

$$\mathrm{Tr}\left(Z^T QZ\right) \geq \alpha$$

for some $\alpha \geq 0$. Such constraint can be incorporated directly into the normalized spectral clustering minimisation problem, and it leads to the following optimisation problem

$$\underset{\substack{U\in\mathbb{R}^{n\times K}\\ U^T DU=I_K\\ \mathrm{Tr}(U^T QU)\geq\alpha}}{\arg\min} \quad \mathrm{Tr}\left(U^T LU\right),$$

which after the change of variable $X = D^{1/2}U$ can be rewritten as

$$\underset{\substack{X\in\mathbb{R}^{n\times K}\\ X^T X=I_K\\ \mathrm{Tr}(X^T \bar{Q}X)\geq\alpha}}{\arg\min} \quad \mathrm{Tr}\left(X^T \mathcal{L}X\right), \tag{5.9}$$

with $\bar{Q} = D^{-1/2}QD^{-1/2}$.

Lemma 5.4. *Let X be solution of* (5.9). *The rows of X are solutions of a generalized eigenvalue problem $\mathcal{L}X_{\cdot k} = \lambda\left(\bar{Q} - \beta\right)X_{\cdot k}$ for some β.*

Proof. The Lagrangian of the minimisation problem (5.9) is

$$\mathrm{Tr}\left(X^T\mathcal{L}X\right) - \lambda\left(\mathrm{Tr}\left(X^T\bar{Q}X\right) - \alpha\right) - \mathrm{Tr}\left(\Gamma^T\left(X^TX - I_K\right)\right),$$

where $\lambda \in \mathbb{R}$ is the Lagrange multiplier associated to the constraint $\mathrm{Tr}\left(X^T\bar{Q}X\right) \geq \alpha$ and $\Gamma \in \mathbb{R}^{K\times K}$ is a symmetric matrix whose elements are the multipliers associated to the constraint $X^TX = I_K$. Note that up to a change of basis, we can choose Γ to be diagonal. Then, according to the KKT Theorem (Kuhn, 1982), any feasible optimal solution of problem (5.9) must verify

$$\begin{aligned} \text{stationarity:} \quad & \mathcal{L}X - \lambda\bar{Q}X - X\Gamma = 0, \\ \text{primal feasibility:} \quad & \operatorname{Tr}\left(X^T\bar{Q}X\right) \geq \alpha \text{ and } X^TX = I_K, \\ \text{dual feasibility:} \quad & \lambda \geq 0, \\ \text{complementary slackness:} \quad & \lambda\left(\operatorname{Tr}\left(X^T\bar{Q}X\right) - \alpha\right) = 0. \end{aligned}$$

The complementary slackness requirement either implies $\lambda = 0$ or $\operatorname{Tr}\left(X^T\bar{Q}X\right) = \alpha$. If $\lambda = 0$, then the stationarity requirement would reduce the problem to the standard (unconstrained) spectral clustering. Thus $\lambda \neq 0$, and $\operatorname{Tr}\left(X^T\bar{Q}X\right) = \alpha$, and the KKT conditions become

$$\begin{aligned} \mathcal{L}X - \lambda\bar{Q}X - X\Gamma &= 0, \\ \operatorname{Tr}\left(X^T\bar{Q}X\right) &= \alpha, \\ X^TX &= I_K, \\ \lambda &> 0. \end{aligned}$$

Since Γ is diagonal, the first equation is equivalent to

$$(\mathcal{L} - \Gamma_{kk})\, X_{\cdot k} = \lambda\bar{Q}X_{\cdot k},$$

which is a generalised eigenvalue problem for a given Γ_{kk}. We end the proof by letting $\beta = -\frac{\Gamma_{kk}}{\lambda}$. □

Based on Lemma 5.4, and following Wang and Davidson, 2010 and Wang *et al.*, 2014, we propose the following procedure:

(i) find the vectors $v_1, \cdots, v_p$ solutions of $\mathcal{L}v_k = \lambda_k\left(\bar{Q} - \beta I_n\right)v_k$ associated to $\lambda_k > 0$;
(ii) given all the feasible eigenvectors $v_1, \cdots, v_p$, pick the top $K - 1$ in terms of minimising $v^T\mathcal{L}v$, and let those $K - 1$ vectors form the columns of X.

Since $\beta < \lambda_K$, there is at least $K-1$ solutions of the generalised eigenvalue problem associated with positive eigenvalues. Furthermore, the solutions are real vectors, since $\mathcal{L}$ and $\bar{Q}-\beta I_n$ are Hermitian matrices. Finally, this procedure is justified, since X verifies the KKT conditions derived in the proof of Lemma 5.4 if $(K-1)\beta < \alpha$. Indeed, since $\mathcal{L}$ is positive semi-definite, we have $v^T\mathcal{L}v \geq 0$ with equality only for $v \propto 1_n$. Hence $\operatorname{Tr}\left(X^T\mathcal{L}X\right) > 0$. Moreover, $\operatorname{Tr}\left(X^T\mathcal{L}X\right) = \sum_{k=0}^{K} X_{\cdot k}^T\mathcal{L}X_{\cdot k} = \sum_k \lambda_k X_{\cdot k}\left(\bar{Q} - \beta I_n\right)X_{\cdot k} \geq \operatorname{Tr}\left(X^T\bar{Q}X\right) - (K-1)\beta$, and $\operatorname{Tr} B < \alpha$. We summarise the procedure in Algorithm 11.

Algorithm 11: Constrained spectral clustering (Wang and Davidson, 2010; Wang *et al.*, 2014).

Input: graph G, must-link/cannot-link matrix Q, number of clusters K, parameter β.

Output: node labelling $\hat{z} \in [K]^n$.

Process: let $\mathcal{L}$ be the normalised Laplacian of G, and let $\bar{Q} = D^{-1/2} Q D^{-1/2}$.

if $\beta \geq \lambda_{K-1}\left(\bar{Q}\right)$ **then**

  Return: $\emptyset$.

else

- let $v_1, \cdots, v_p$ be solution of the generalised eigenvalue problem $\mathcal{L}v = \lambda\left(\bar{Q} - \beta\right) v$ associated with eigenvalues $\lambda > 0$;
- let $V^* = \arg\min_{V \in \mathbb{R}^{n \times K-1}} \operatorname{Tr}\left(V^T \mathcal{L} V\right)$ where the columns of V are a subset of the feasible eigenvectors computed previously.

  Return: $\hat{z} = \text{k-means}(D^{-1/2} V^*, K)$

5.3.2 Laplacian Regularization

Previous methods minimise a cost function, which involves a smoothness term $\operatorname{Tr}(X^T M X)$, where M is typically the graph (standard or normalised) Laplacian, a penalty term penalising any differences between $X_{\ell \cdot}$ and $S_{\ell \cdot}$, and eventually a regularisation term.

In contrast, *Laplacian regularization* (Belkin and Niyogi, 2002) enforces the smoothness by constraining the vector X to belong to the eigenspace of the graph Laplacian L spanned by the eigenvectors associated to the p smallest eigenvalues. It then finds the linear combination of these eigenvectors that minimises the mean-squared error between X and S on the labelled nodes.

Let $v_1, \ldots, v_p$ be the eigenvectors of L associated to the p smallest eigenvalues, normalized so that $\|v_i\|_2^2 = 1$. The solution $X = (x_{ik})_{i \in [n], k \in [K]}$ is written as $x_{ik} = \sum_{q=1}^{p} b_{qk} v_q(i)$ where $v_q(i)$ stands for the i-th entry of the eigenvector v_q. In matrix form, this gives $X = VB$ where $V = (v_1, \ldots, v_p)$ and $B \in \mathbb{R}^{p \times K}$. The mean-squared error between the labelled nodes and their corresponding oracle value is then

$$\operatorname{MSE}\left(X_{\ell \cdot}, S_{\ell \cdot}\right) = \sum_{k=1}^{K} \sum_{i \in \ell} (x_{ik} - s_{ik})^2 = \sum_{k=1}^{K} \sum_{i \in \ell} \left(\sum_{q=1}^{p} b_{qk} v_{qi} - s_{ik} \right)^2 .$$

Let $\tilde{b} = b_{\cdot k}$ and $\tilde{s} = S_{\cdot k}$ be the k-th columns of B and S, respectively. The solution to the least square problem

$$\underset{\tilde{b}\in\mathbb{R}^p}{\arg\min} \sum_{i\in\ell} \left(\sum_{q=1}^{p} \tilde{b}_q v_{qi} - \tilde{s}_i \right)^2$$

tis given by $\tilde{b} = \left(V_{\ell\cdot}^T V_{\ell\cdot}\right)^{-1} V_{\ell\cdot}\tilde{s}_\ell$. Therefore, the solution $\widehat{X}^{LR}$ minimising the mean-squared error is

$$\widehat{X}^{LR} = \left(V_{\ell\cdot}^T V_{\ell\cdot}\right)^{-1} V_{\ell\cdot}\tilde{s}_\ell.$$

This is summarised in Algorithm 12.

Algorithm 12: Laplacian regularization (Belkin and Niyogi, 2002).

Input: graph G, oracle S, number of eigenvectors p.
Output: node labelling $\hat{z} \in [K]^n$.
Process:

- compute $v_1, \ldots, v_p$ the orthonormal eigenvectors associated to the p smallest eigenvalues of the graph Laplacian $L = D - A$;
- let $V = (v_1 \ldots, v_p) \in R^{n\times p}$;
- let $\widehat{X}^{LR} = V\widehat{B}^{LR}$ where $\widehat{B}^{LR}$ is the solution of $(I_\ell V)\,\widehat{B}^{LR} = S$;
- for $i = 1, \ldots, n$ let $\hat{z}_i$ be defined using the classification rule (5.2) on $\widehat{X}^{LR}$.

Return: $\hat{z}$.

5.3.3 ℓ^1-based Methods: Sparse Label Propagation

Previous methods consist in minimising a cost function based on the ℓ^2-norm. Instead, Jung *et al.*, 2019 proposed to measure the smoothness of a signal $x \in \mathbb{R}^n$ on a graph via its total variation

$$\|x\|_{\mathrm{TV}} = \sum_{i,j} a_{ij} \left|x_i - x_j\right|.$$

If we let $z^0 \in [K]^n$ be the community labels, and ℓ be the set of labelled nodes, then one can state the following optimisation problem

$$\hat{x} = \underset{\substack{x\in\mathbb{R}^n \\ \forall i\in\ell:\, x_i = z_i^0}}{\arg\min} \sum_{i,j} a_{ij} \left|x_i - x_j\right|. \tag{5.10}$$

We can recover the predicted communities $\hat{z}$ by truncating $\hat{x} \in \mathbb{R}^n$ to $\hat{z} \in [K]^n$.

We recall that the standard Label Propagation (5.3) consists in minimising $x^T Lx = \sum_{i,j} a_{ij} \left(x_i - x_j\right)^2$ under the oracle constraints. Hence, problem (5.10) resembles Label Propagation, except that it involves the ℓ^1-norm of signal differences along the graph edges. Therefore, we expect it to accurately learn signals which abruptly vary over few edges (which is indeed the case of community labels). In contrast, an ℓ^2-norm based methods like Label Propagation might smooth out such abrupt variations.

Finally, since the optimisation problem (5.10) involves nondifferentiable function, it makes the theoretical analysis harder and rules out some popular methods like gradient-based ones. We refer the reader to Jung *et al.*, 2019 for the theoretical analysis and for the details of algorithmic implementation, and we simply state Algorithm 13.

Algorithm 13: Sparse Label Propagation (Jung *et al.*, 2019).

Input: graph $G = (V, E)$, labelled set ℓ, initial labels $\left(z_i^0\right)_{i \in \ell}$ and number of iterations $n_{\text{iterations}}$.

Output: predicted node labelling $\hat{z} \in [K]^n$.

Initialize: let $k = 0$, $z^{(0)} = z_\ell$, $\hat{z}^{(0)} = 0_n$, $\hat{y}^{(0)} = 0_n$, $\gamma_i = \frac{1}{\sum_{j \in \mathcal{N}_i} A_{ij}}$ for $i \in [n]$ and $\lambda_{(ij)} = \frac{1}{2A_{ij}}$ for $(ij) \in E$. Define $\Gamma = \text{diag}(\gamma)$, $\Lambda = \text{diag}\left(\lambda_{(ij)}\right)_{(ij) \in E}$ and $\mathcal{I} \in \{0, 1\}^{|E| \times n}$ the incidence matrix of G.

Update: **while** $k < n_{\text{iterations}}$ **do**

- $z^{(k+1)} = z^{(k)} - \Gamma I y$;
- $z_i^{(k+1)} = z_i^0$ for all $i \in \ell$;
- $y = y + \Lambda \mathcal{I}^T \left(2z^{(k+1)} - z^{(k)}\right)$;
- $y_{(ij)} = \frac{y_{(ij)}}{\max\{1, y_{ij}\}}$ for all edge $(ij) \in E$;
- $\hat{z} = \left(1 - \frac{1}{k+1}\right) \hat{z} + \frac{1}{k+1} z^{(k+1)}$;
- $k = k + 1$.

Return: $\hat{z}$.

5.4 Bayesian Approach to SSL and Its Theoretical Analysis

This section studies theoretical properties of Bayesian estimators for DC-SBM graphs in the SSL setting. For the simplicity of exposition, we mostly consider the case of $K = 2$ clusters. The prior latent block structure is given by a random vector $z^0 = \left(z_1^0, \ldots, z_n^0\right)$ with $z_i^0 \sim \mathrm{Uni}\left(\{-1, 1\}\right)$. The oracle is then represented as a vector $s \in \{0, -1, 1\}^n$, whose entries s_i are independent and distributed as follows:

$$s_i = \begin{cases} z_i^0, & \text{with probability} \quad \eta_1, \\ -z_i^0, & \text{with probability} \quad \eta_0, \\ 0, & \text{otherwise.} \end{cases} \tag{5.11}$$

5.4.1 MAP Estimator for DC-SBM with a Noisy Oracle

Proposition 5.1 (Adapted from Avrachenkov and Dreveton, 2020)**.** *Let A be the adjacency matrix of a homogeneous Poisson SBM as in* (2.7)*, with $\omega_{\mathrm{in}} > \omega_{\mathrm{out}}$ and s be an oracle information, defined by* (5.11)*. The* Maximum A Posteriori *(MAP) estimator of the true class labelling is given by*

$$\hat{z}_{\mathrm{MAP}} = \arg\max_{z \in [K]^n} \mathbb{P}\left(z \,|\, A, s\right),$$

which is equal to

$$\arg\min_{z \in [K]^n} \mathrm{Cut}(A, z) - \tau n_1(z) n_2(z) + \lambda \left|i \in \ell : z_i \neq s_i\right|, \tag{5.12}$$

where $\tau = \frac{\omega_{\mathrm{in}} - \omega_{\mathrm{out}}}{\log \frac{\omega_{\mathrm{in}}}{\omega_{\mathrm{out}}}}$, $\lambda = \frac{\log \frac{\eta_1}{\eta_0}}{\log \frac{\omega_{\mathrm{in}}}{\omega_{\mathrm{out}}}}$ and $n_k(z) = \sum_{i=1}^n 1(z_i = k)$ is the number of nodes assigned to class k by labelling z.

The term $n_1(z) n_2(z) = n_1(z)(n - n_1(z))$ is maximal when $n_1(z) = \frac{n}{2}$, *i.e.*, when z predicts two clusters of same size. Therefore, the MAP estimator in the SSL context shows a trade-off between two unsupervised terms (minimising the graph's cut and having balanced community sizes) and a semi-supervised term (minimising the number of disagreements between the oracle and the prediction).

Proof. First, we have from Bayes formula

$$\mathbb{P}\left(z \,|\, A, s\right) \propto \mathbb{P}\left(A \,|\, s, z\right) \mathbb{P}\left(z \,|\, s\right),$$

where the proportionality hides a term $\mathbb{P}\left(A \,|\, s\right)$ independent of z. We established at the end of the proof of Proposition 4.4 that

$$\log \mathbb{P}\left(A \,|\, z\right) = \frac{1}{2} \log \frac{\omega_{\mathrm{in}}}{\omega_{\mathrm{out}}} \sum_{i \neq j} \left(A_{ij} - \frac{\omega_{\mathrm{in}} - \omega_{\mathrm{out}}}{\log \frac{\omega_{\mathrm{in}}}{\omega_{\mathrm{out}}}} \theta_i \theta_j \right) 1\left(z_i = z_j\right) + C,$$

where C is a constant independent of z. Finally, the oracle information, given by the term $\mathbb{P}(z\,|\,s)$, is equal to

$$
\begin{aligned}
\mathbb{P}(z\,|\,s) &= \prod_{i=1}^{n} \frac{\mathbb{P}(s_i\,|\,z_i)}{\mathbb{P}(s_i)} \mathbb{P}(z_i) \\
&= \left(\frac{\eta_1}{\eta_1+\eta_0}\right)^{\left|\{i\in\ell:\, z_i=s_i\}\right|} \left(\frac{\eta_0}{\eta_1+\eta_0}\right)^{\left|\{i\in\ell:\, z_i\neq s_i\}\right|} \left(\frac{1}{2}\right)^{n} \\
&= \left(\frac{\eta_0}{\eta_1}\right)^{\left|\{i\in\ell\,:\, z_i\neq s_i\}\right|} \left(\frac{\eta_1}{\eta_1+\eta_0}\right)^{\left|\ell\right|} \left(\frac{1}{2}\right)^{n}, \qquad (5.13)
\end{aligned}
$$

where we used $\left|\{i \in \ell\colon z_i = s_i\}\right| + \left|\{i \in \ell\colon z_i \neq s_i\}\right| = \left|\ell\right|$ in the last line. □

5.4.2 Continuous Relaxation

The MAP estimator derived in Proposition 5.1 can be rewritten as

$$
\hat{z}^{\mathrm{MAP}} = \underset{z\in\{-1,1\}^n}{\arg\min} \; -z^T\left(A - \tau 1_n^T 1_n\right) z + \lambda\,(s-\mathcal{P}z)^T\,(s-\mathcal{P}z)\,,
$$

where $\mathcal{P}$ is the diagonal matrix whose element (i, i) is equal to 1, if $i \in \ell$, and it is equal to 0, otherwise. First, we simply notice that

$$
|i \in \ell\colon z_i \neq s_i| = \frac{1}{4}\sum_{i\in\ell}(s_i - z_i)^2 = \frac{1}{4}(s-\mathcal{P}z)^T\,(s-\mathcal{P}z)\,.
$$

We then perform a continuous relaxation mirroring what is commonly done for unsupervised spectral methods (Newman, 2013) and discussed in Section 4.4.2, namely, we consider the following optimisation problem

$$
\widehat{X} = \underset{\substack{x\in\mathbb{R}^n \\ \sum_i \kappa_i x_i^2 = \sum_i \kappa_i}}{\arg\min} \left(-x^T A_\tau x + \lambda (s-\mathcal{P}x)^T (s-\mathcal{P}x)\right), \qquad (5.14)
$$

where $A_\tau = A - \tau 1_n 1_n^T$ and $\kappa = (\kappa_1, \ldots, \kappa_n)$ is a vector of positive entries. For the simplicity of the derivations, we choose to constrain x to the hyper-sphere $\|x\|^2 = n$ by letting $\kappa_i = 1$, but other choices would lead to a similar analysis. In particular, in the numerical Section 5.4.4 we will compare this choice with a degree-normalization approach ($\kappa_i = d_i$).

We further note that for the perfect oracle the corresponding relaxation is

$$
\widehat{X} = \underset{\substack{x\in\mathbb{R}^n \\ x_\ell = s_\ell \\ \|x\|^2=n}}{\arg\min} \left(-x^T A_\tau x\right). \qquad (5.15)
$$

Given the classification vector $\widehat{X} \in \mathbb{R}^n$, node i is classified into cluster $\hat{z}_i \in \{-1, 1\}$ such that

$$\hat{z}_i = \begin{cases} 1 & \text{if} \quad \widehat{X}_i > 0, \\ -1 & \text{otherwise.} \end{cases} \tag{5.16}$$

Let us solve the minimisation problem (5.14). By letting $\gamma \in \mathbb{R}$ be the Lagrange multiplier associated with the constraint $\|x\|^2 = n$, the Lagrangian of the optimisation problem (5.14) is then

$$-x^T A_\tau x + \lambda (s - \mathcal{P}x)^T (s - \mathcal{P}x) - \gamma \left(x^T x - n \right).$$

This leads to the *constrained* linear system

$$\begin{cases} (-A_\tau + \lambda \mathcal{P} - \gamma I_n) x = \lambda s, \\ x^T x = n, \end{cases} \tag{5.17}$$

whose unknowns are γ and x.

The exact optimal value of γ can be found explicitly following Gander *et al.*, 1989. Firstly, we note that if (γ_1, x_1) and (γ_2, x_2) are solutions of the system (5.17), then

$$\mathcal{C}(x_1) - \mathcal{C}(x_2) = \frac{\gamma_1 - \gamma_2}{2} \|x_1 - x_2\|^2,$$

where $\mathcal{C}(x) = -x^T A_\tau x + \lambda (s - \mathcal{P}x)^T (s - \mathcal{P}x)$ is the cost function minimised in (5.14). Hence, among the solution pairs (γ, x) of the system (5.17), the solution of the minimisation problem (5.14) is the vector x associated with the smallest γ.

Secondly, the eigenvalue decomposition of $-A_\tau + \lambda \mathcal{P}$ reads as

$$-A_\tau + \lambda \mathcal{P} = Q \Delta Q^T,$$

where $\Delta = \text{diag}(\delta_1, \ldots, \delta_n)$ with $\delta_1 \leq \cdots \leq \delta_n$ and $Q^T Q = I_n$. Therefore, after the change of variables $u = Q^T x$ and $b = \lambda Q^T s$, the system (5.17) is transformed to

$$\begin{cases} \Delta u = \gamma u + b, \\ u^T u = n. \end{cases}$$

Thus, the solution $\widehat{X}$ of the optimisation problem (5.14) verifies

$$(-A_\tau + \lambda \mathcal{P} - \gamma_* I_n) \widehat{X} = \lambda s, \tag{5.18}$$

where γ_* is the smallest solution of the *explicit secular equation* (Gander *et al.*, 1989)

$$\sum_{i=1}^{n} \left(\frac{b_i}{\delta_i - \gamma} \right)^2 - n = 0. \tag{5.19}$$

We summarise this in Algorithm 14. Note that for the sake of generality, we let λ and τ be hyper-parameters of the algorithm. If the model parameters are known, we can use the expressions of λ and τ derived in Proposition 5.1. The choice of λ and τ is further discussed in Section 5.4.4.

Algorithm 14: Semi-supervised learning by a MAP relaxation.

Input: adjacency matrix A, oracle information s, parameters τ and λ.
Output: node labelling $\hat{z} \in [K]^n = (\hat{z}_1, \ldots, \hat{z}_n)$.
Process:

- let γ^* be the smallest solution of equation (5.19);
- compute $\widehat{X}$ as the solution of equation (5.18);
- for $i = 1, \ldots, n$ let $\hat{z}_i$ be defined using (5.16) on $\widehat{X}$.

Return: $\hat{z}$.

5.4.3 Upper Bound on the Number of Misclassified Nodes

In this section, we derive an upper bound on the number of unlabelled nodes misclassified by Algorithm 14 on a DC-SBM. We then specialise the results for some particular cases. We will assume that given $(p_{\text{in}}, p_{\text{out}}, \theta, z)$, the graph adjacency matrix $A = (a_{ij})$ is generated as

$$a_{ij} = a_{ji} \sim \begin{cases} \text{Ber}\left(\theta_i \theta_j p_{\text{in}}\right), & \text{if } \; z_i = z_j, \\ \text{Ber}\left(\theta_i \theta_j p_{\text{out}}\right), & \text{otherwise,} \end{cases} \tag{5.20}$$

for $i < j$, and $A_{ii} = 0$. Furthermore, we suppose that $z_i \sim \text{Uni}\left(\{-1, 1\}\right)$, and that the entries of θ are independent random variables satisfying $\theta_i \in [\theta_{\min}, \theta_{\max}]$ with $\mathbb{E}\theta_i = 1$, $\theta_{\min} > 0$, and $\theta_{\max}^2 \max(p_{\text{in}}, p_{\text{out}}) \le 1$.

For an estimator $\hat{z} \in \{-1, 1\}^n$ of z, the number of mis-clustered nodes is simply the Hamming distance between the two sequences $\hat{z}$ and z, defined as

$$d_{\text{Ham}}(\hat{z}, z) = \sum_{i=1}^{n} 1\left(\hat{z}_i \neq z_i\right),$$

and the proportion of mis-clustered nodes is $\frac{d_{\text{Ham}}(\hat{z}, z)}{n}$. Note that, unlike in the unsupervised clustering, we do not take a minimum over the permutations of the predicted labels since we should be able to learn the correct community labels from the informative oracle.

Theorem 5.5 (Avrachenkov and Dreveton, 2020). *Consider a DC-SBM with a noisy oracle as defined in* (5.20),(5.11). *Let* $\bar{d} = \frac{n}{2}(p_{\text{in}} + p_{\text{out}})$ *and* $\bar{\alpha} = \frac{n}{2}(p_{\text{in}} - p_{\text{out}})$. *Suppose that* $\tau > p_{\text{out}}$, *and let* $\hat{z}$ *be the output of Algorithm 14. Then, the proportion of misclassified unlabelled nodes verifies*

$$\frac{d_{\text{Ham}}(\hat{z}_u, z_u)}{n} \leq C\left(\frac{p_{\text{in}} + p_{\text{out}}}{p_{\text{in}} - p_{\text{out}}}\right)^2 \left(\frac{\bar{\alpha} + \lambda}{\lambda}\right)^2 \frac{1}{(\eta_1 + \eta_0)(\eta_1 - \eta_0)^2 \bar{d}}.$$

In the following, the mean-field graph refers to the weighted graph formed by the expected adjacency matrix of a DC-SBM graph. Moreover, we assume without loss of generality that the first $\frac{n}{2}$ nodes are in the first cluster and the last $\frac{n}{2}$ are in the second cluster. Therefore, $\mathbb{E}A = ZBZ^T$ with $B = \begin{pmatrix} p_{\text{in}} & p_{\text{out}} \\ p_{\text{out}} & p_{\text{in}} \end{pmatrix}$ and $Z = \begin{pmatrix} 1_{n/2} & 0_{n/2} \\ 0_{n/2} & 1_{n/2} \end{pmatrix}$. In particular, the coefficients θ_i disappear because $\mathbb{E}\theta_i = 1$. We consider the setting where diagonal elements of $\mathbb{E}A$ are not zeros. This accounts for modifying the definition of DC-SBM, where we can have self-loops with probability p_{in}. Nonetheless, we could set the diagonal elements of $\mathbb{E}A$ to zeros and our results would still hold at the expense of cumbersome expressions. Note that the matrix $\mathbb{E}A$ has two non-zero eigenvalues: $\bar{d} = n\frac{p_{\text{in}}+p_{\text{out}}}{2}$ and $\bar{\alpha} = n\frac{p_{\text{in}}-p_{\text{out}}}{2}$.

Proof of Theorem 5.5. We prove the statement in three steps. We first show that the solution $\widehat{X}$ of the constrained linear system (5.17) is concentrated around the solution $\bar{x}$ of the same system for the mean-field model. Then, we compute $\bar{x}$ and show that we can retrieve the correct cluster assignment from it. We finally conclude with the derivation of the bound.

(i) Similarly to (Avrachenkov *et al.*, 2018c) and (Avrachenkov and Dreveton, 2019), let us rewrite equation (5.18) as a perturbation of a system of linear equations corresponding to the mean-field solution. We thus have

$$\left(\mathbb{E}\tilde{\mathcal{L}} + \Delta\tilde{\mathcal{L}}\right)\left(\bar{x} + \Delta x\right) = \lambda s,$$

where $\tilde{\mathcal{L}} = -A_\tau + \lambda\mathcal{P} - \gamma_* I_n$, $\Delta x := \widehat{X} - \bar{x}$ and $\Delta\tilde{\mathcal{L}} := \tilde{\mathcal{L}} - \mathbb{E}\tilde{\mathcal{L}}$.

A perturbation of a system of linear equations $(A + \Delta A)(x + \Delta x) = b$ leads to the following sensitivity inequality (Horn and Johnson, 2012, Section 5.8):

$$\frac{\|\Delta x\|}{\|x\|} \leq \kappa(A)\frac{\|\Delta A\|}{\|A\|},$$

where $\|.\|$ is the operator norm associated to a vector norm $\|.\|$ (we use the same notations for simplicity) and $\kappa(A) := \|A^{-1}\| \cdot \|A\|$ is the condition number. In our case, the above inequality can be rewritten as follows:

$$\frac{\|\widehat{X} - \bar{x}\|}{\|\bar{x}\|} \le \left\| \left(\mathbb{E}\, \tilde{\mathcal{L}} \right)^{-1} \right\| \cdot \left\| \Delta\, \tilde{\mathcal{L}} \right\|, \tag{5.21}$$

employing the Euclidean vector norm and the spectral operator norm. The spectral study of $\mathbb{E}\, \tilde{\mathcal{L}}$ (see Corollary B.2 in Appendix B.1.1) gives:

$$\left\| \left(\mathbb{E}\, \tilde{\mathcal{L}} \right)^{-1} \right\| = \frac{1}{\min\left\{ |\lambda| : \lambda \in \mathrm{Sp}\left(\mathbb{E}\, \tilde{\mathcal{L}} \right) \right\}} = \frac{1}{-t_2^+ - \bar{\gamma}_*},$$

where t_2^+ is defined in Corollary B.2 in Appendix B.1.1 and $\bar{\gamma}_*$ is the solution of equation (5.19) for the mean-field model. Lemma B.3 in Appendix B.1.2 leads to

$$\left\| \left(\mathbb{E}\, \tilde{\mathcal{L}} \right)^{-1} \right\| \le \frac{1}{\lambda + \bar{\alpha}}. \tag{5.22}$$

The last ingredient needed is the concentration of the adjacency matrix around its expectation. We have

$$\left\| \tilde{\mathcal{L}} - \mathbb{E}\tilde{\mathcal{L}} \right\| \le \|(\gamma_* - \bar{\gamma}_*)\, I_n\| + \|A - \mathbb{E}\, A\| \le |\, \gamma_* - \bar{\gamma}_* \,| + \|A - \mathbb{E}\, A\|.$$

Proposition B.2 in Appendix B.1.3 shows that

$$|\, \gamma_* - \bar{\gamma}_* \,| \le \left(1 + \frac{27\,(\bar{\alpha} + \lambda)^3}{\sqrt{2}\sqrt{\eta_1 + \eta_0}(\eta_1 - \eta_0)\bar{\alpha}^2\lambda} \right) \sqrt{\bar{d}}.$$

Moreover, when $d = \Omega(\log n)$, we have $\|A - \mathbb{E}\, A\| = O\left(\sqrt{\bar{d}}\right)$ (Feige and Ofek, 2005). If $\bar{d} = o(\log n)$, the same result holds with a proper pre-processing on A, and we refer the reader to (Le *et al.*, 2017) for more details. To keep notations short, we will omit this extra step in the proof. Using this concentration bound, we have

$$\begin{aligned}
\left\| \tilde{\mathcal{L}} - \mathbb{E}\tilde{\mathcal{L}} \right\| &\le \left(C' + \frac{27\,(\bar{\alpha} + \lambda)^3}{\sqrt{2}\sqrt{\eta_1 + \eta_0}(\eta_1 - \eta_0)\bar{\alpha}^2\lambda} \right) \sqrt{\bar{d}} \\
&\le \left(C' + \frac{27}{\sqrt{2}} \right) \frac{(\lambda + \bar{\alpha})^3}{\bar{\alpha}^2\lambda} \frac{\sqrt{\bar{d}}}{\sqrt{\eta_1 + \eta_0}\,(\eta_1 - \eta_0)}
\end{aligned}$$

for some constant C'. Let $C = C' + \frac{27}{\sqrt{2}}$. By combining the above with inequality (5.22), the inequality (5.21) becomes

$$\frac{\|\widehat{X} - \bar{x}\|}{\|\bar{x}\|} \leq C \frac{(\lambda + \bar{\alpha})^2}{\bar{\alpha}^2 \lambda} \frac{\sqrt{\bar{d}}}{\sqrt{\eta_1 + \eta_0}\,(\eta_1 - \eta_0)}. \tag{5.23}$$

(ii) Node i in the mean-field model is correctly classified by decision rule (5.16) if the sign of $\bar{x}_i$ equals the sign of z_i. Corollary B.5 in Appendix B.2 shows that this is indeed the case for the unlabelled nodes.

(iii) Finally, for an unlabelled node i to be correctly classified, the node's value $\widehat{X}_i$ should be close enough to its mean-field value $\bar{x}_i$. In particular, the part (ii) shows that if $|\widehat{X}_i - \bar{x}_i|$ is smaller than some non-vanishing constant β, then an unlabelled node i will be correctly classified. An unlabelled node i is said to be β-bad if $\left|\widehat{X}_i - \bar{x}_i\right| > \beta$. We denote by S_β the set of β-bad nodes. The nodes that are not β-bad are a.s. correctly classified, and thus $d_{\text{Ham}}(\hat{z}_u, z_u) \leq |S_\beta|$. From $\left\|\widehat{X} - \bar{x}\right\|^2 \geq \sum_{i \in S_\beta} \left|\widehat{X}_i - \bar{x}_i\right|^2$, it follows that $\left\|\widehat{X} - \bar{x}\right\|^2 \geq \beta^2 |S_\beta|$. Thus, using (5.23) and the norm constraint $\|\bar{x}\|^2 = n$, we have

$$\left|S_\beta\right| \leq \frac{1}{\beta^2} \left(\frac{C}{\eta_1 - \eta_0} \frac{\bar{\alpha} + \lambda}{\bar{\alpha}\lambda} \sqrt{\bar{d}} \right)^2 n,$$

for some constant C. We end the proof by noticing that $\frac{\bar{d}}{\bar{\alpha}} = \frac{p_{\text{in}} + p_{\text{out}}}{p_{\text{in}} - p_{\text{out}}}$. □

Corollary 5.6 (Almost exact recovery in the diverging degree regime). *Consider a DC-SBM such that* $\bar{d} \gg 1$, $\frac{p_{\text{in}} + p_{\text{out}}}{p_{\text{in}} - p_{\text{out}}} = O(1)$, *and* $\sqrt{\eta_0 + \eta_1}(\eta_1 - \eta_0) \gg \frac{1}{\sqrt{\bar{d}}}$. *Suppose that* $\tau > p_{\text{out}}$ *and* $\lambda \gtrsim \bar{\alpha}$. *Then, Algorithm 14 correctly classifies almost all the unlabelled nodes.*

Proof. With the corollary's assumptions $(\eta_1 - \eta_0)^2 \bar{d} \to +\infty$ and $\frac{\bar{\alpha}+\lambda}{\lambda} = O(1)$. Thus, by Theorem 5.5 the fraction of misclassified nodes is $o(1)$. □

The quantity $(\eta_1 - \eta_0)n$ is the expected difference between the number of nodes correctly labelled and the number of nodes wrongly labelled by the oracle. In particular, Corollary 5.6 allows for a sub-linear number of labelled nodes since η_0 and η_1 can go to zero.

Corollary 5.7 (Detection in the constant degree regime). *Consider a DC-SBM such that* $p_{\text{in}} = \frac{c_{\text{in}}}{n}$ *and* $p_{\text{out}} = \frac{c_{\text{out}}}{n}$ *where* $c_{\text{in}}, c_{\text{out}}$ *are constants. Suppose that* $\sqrt{\eta_0 + \eta_1}(\eta_1 - \eta_0)$ *is a non-zero constant, and let* $\tau > 2p_{\text{out}}$ *and* $\lambda \gtrsim 1$. *Then, for* $\frac{(c_{\text{in}} - c_{\text{out}})^2}{c_{\text{in}} + c_{\text{out}}}$ *bigger than some constant, w.h.p. Algorithm 14 performs better than a random guess.*

Proof. According to Theorem 5.5, the fraction of misclustered nodes is smaller than $\frac{1}{2}$ when $\frac{(c_{\text{in}}-c_{\text{out}})^2}{c_{\text{in}}+c_{\text{out}}}$ is larger than $\frac{2C}{(\eta_1-\eta_0)^2}\left(\frac{\bar{a}+\lambda}{\lambda}\right)^2$, which is lower bounded by a constant. □

The quantity $\frac{(c_{\text{in}}-c_{\text{out}})^2}{c_{\text{in}}+c_{\text{out}}}$ can be interpreted as the signal-to-noise ratio. It is unfortunate that Corollary 5.7 does not allow us to control the constant in the statement of the corollary. This constant comes from concentration of the adjacency matrix. Similar remarks were made in (Le *et al.*, 2017) for the analysis of unsupervised spectral clustering in the constant degree regime for SBM graphs.

5.4.4 Numerical Results

This section presents numerical experiments both on synthetic data sets generated from DC-SBMs and on real networks. In particular, we discuss the impact of the oracle mistakes (defined by the ratio $\frac{\eta_0}{\eta_0+\eta_1}$) on the performance of the algorithms.

Choice of λ and τ Let us denote by σ_1 and σ_2 the largest and second largest eigenvalues of A. We choose $\tau = \frac{4}{n}(\sigma_1+\sigma_2)$ and $\lambda = \frac{\log\frac{\eta_1}{\eta_0}}{\log\frac{\sigma_1+\sigma_2}{\sigma_1-\sigma_2}}$ if $\eta_0 \neq 0$, and $\lambda = \frac{\log(n\eta_1)}{\log\frac{\sigma_1+\sigma_2}{\sigma_1-\sigma_2}}$ otherwise. The heuristic for this choice is as follows. For a SBM graph, we have $\sigma_1 \approx \frac{n}{2}\left(p_{\text{in}}+p_{\text{out}}\right)$ and $\sigma_2 \approx \frac{n}{2}(p_{\text{in}}-p_{\text{out}})$, hence $\frac{4}{n}(\sigma_1+\sigma_2) = 2p_{\text{in}} > p_{\text{out}}$, and τ verifies the condition of Theorem 5.5. For λ, we have $\frac{\log\frac{\eta_1}{\eta_0}}{\log\frac{\sigma_1+\sigma_2}{\sigma_1-\sigma_2}} \approx \frac{\log\frac{\eta_1}{\eta_0}}{\log\frac{p_{\text{in}}}{p_{\text{out}}}}$, which is indeed close to the expression of λ derived in Proposition 5.1 if $p_{\text{in}}, p_{\text{out}} = o(1)$.

Choice of relaxation We first compare the choice of the constraint in the continuous relaxation (5.14). Specifically, we compare the choice $\sum_i x_i^2 = n$ (refered to as *standard relaxation*) versus $\sum_i d_i x_i^2 = 2|E|$ (refered to as *degree-normalized relaxation*). This leads to two versions of Algorithm 14, whose cost obtained on SBM graph with a noisy oracle is presented in Figure 5.5. In particular, we observe that the normalized choice leads to a smaller cost. Therefore, in the following we will only consider the version of Algorithm 14 solving the relaxed problem (5.14) with constraint $\sum_i d_i x_i^2 = 2|E|$ instead of $\sum_i x_i^2 = n$, as it gives better numerical results.

Experiments on synthetic graphs We first consider clustering on DC-SBM. We set $n = 2000$, $p_{\text{in}} = 0.04$ and $p_{\text{out}} = 0.02$. We consider three scenarios:

- In Figure 5.6(a) we consider a standard SBM ($\theta_i = 1$ for all i).

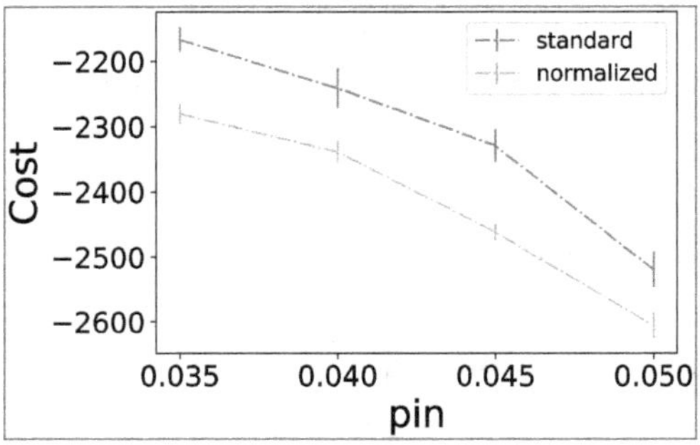

Figure 5.5. Cost in Algorithm 14 with the standard and degree-normalized versions of the constraint, on 50 realizations of SBM with $n = 500$, $p_{\text{out}} = 0.03$ and 50 labelled nodes with 10% noise.

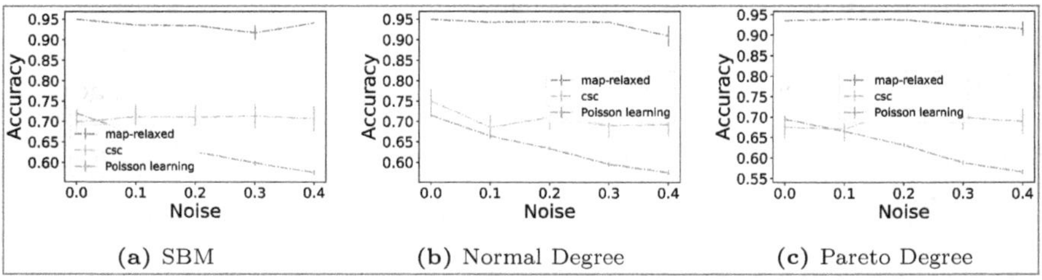

Figure 5.6. Average accuracy obtained by different semi-supervised clustering methods on DC-SBM graphs, with $n = 1000$, $p_{\text{in}} = 0.04$ and $p_{\text{out}} = 0.02$ with different distributions for θ. The number of labelled nodes is equal to 40. Accuracies are computed on the unlabelled nodes and are averaged over 50 realisations; the error bars show the standard error.

- In Figure 5.6(b) we generate θ_i according to $|\mathcal{N}(0, \sigma^2)| + 1 - \sigma\sqrt{2/\pi}$ where $|\mathcal{N}(0, \sigma^2)|$ denotes the absolute value of a normal random variable with mean 0 and variance σ^2. We take $\sigma = 0.25$. Note that this definition enforces $\mathbb{E}\theta_i = 1$.
- In Figure 5.6(c) we generate θ_i from Pareto distribution with density function $f(x) = \frac{am^a}{x^{a+1}} 1(x \geq m)$ with $a = 3$ and $m = 2/3$ (chosen such that $\mathbb{E}\theta_i = 1$).

We compare the performance of Algorithm 14 (called *map-relaxed* in the figures) with *Poisson learning* (Algorithm 10) and *constrained spectral clustering* (Algorithm 11, abbreviated as *csc*). Results are shown in Figure 5.6. While *map-relaxed* and *csc* limit the decrease of accuracy when the noise increase, the performance of *csc* is quite poor on those synthetic data sets. Furthermore, we notice that *Poisson learning* also gives poor result on the synthetic data sets, and its performance further deteriorates with noise.

Experiments on MNIST data set As a real-life example, we perform simulations on the standard MNIST data set (LeCun *et al.*, 1998). As preprocessing,

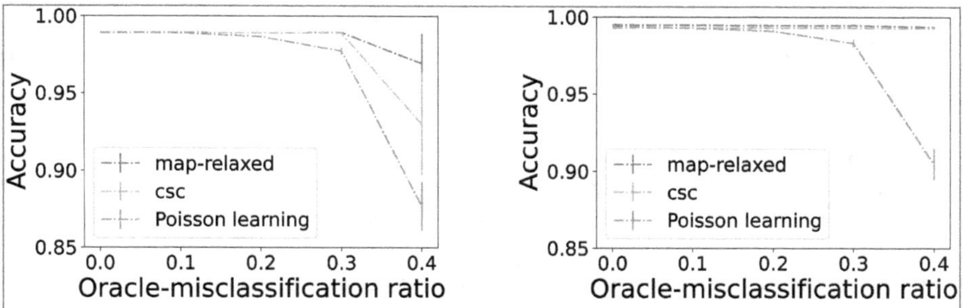

(a) Digits (2,4). (b) Digits (7,8).

Figure 5.7. Average accuracy obtained on a subset of the MNIST data set by different semi-supervised algorithms as a function of the oracle-misclassification ratio, when the number of labelled nodes is equal to 10. Accuracy is averaged over 50 random realizations, and the error bars show the standard error.

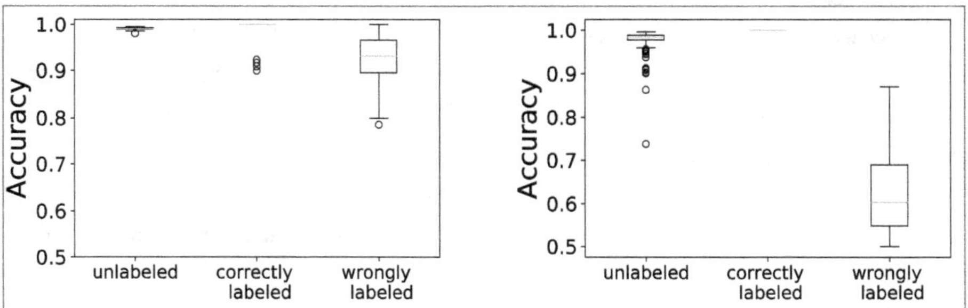

(a) Algorithm 14. (b) *Poisson learning.*

Figure 5.8. Average accuracy obtained on the unlabeled, correctly labeled, and wrongly labelled nodes by the oracle. Simulations are done on 1000 digits (2,4). The noisy oracle correctly classifies 24 nodes and misclassifies 16 nodes, and the boxplots show 100 realizations.

we select 1000 images corresponding to two digits and compute the k-nearest-neighbors graph (we take $k = 8$) with gaussian weights $w_{ij} = \exp\left(-\|x_i - x_j\|^2/s_i^2\right)$ where x_i represents the data for image i and s_i is the average distance between x_i and its K-nearest neighbors. Accuracy for different digit pairs is given in Figure 5.7. We notice that the performance of the three algorithms are excellent. But, under large oracle noise, the accuracy of *Poisson learning* decreases more than the accuracy of Algorithm 14 or *constrained spectral clustering*.

To further highlight the influence of the noise, we plot in Figure 5.8 the accuracy obtained by the three algorithms on the unlabelled nodes, the correctly labelled nodes, and the wrongly labelled nodes. While the accuracy of *Poisson learning* is excellent on the unlabelled nodes, it fails at correctly classifying the wrongly labelled nodes. On the contrary, Algorithm 14 allows for a smoother recovery: the unlabeled, correctly labeled, and wrongly labelled nodes have roughly the same classification accuracy. While some correctly labelled nodes are misclassified, many wrongly labelled nodes become correctly classified, and the unlabelled nodes are better recovered.

Further Notes

In many networks, such as social networks, citation networks and knowledge graphs, the nodes have features. Thus, it is very natural to try to take into account both the graph structure and the features. This idea has been implemented in *Graph Neural Networks (GNNs)*. Scarselli *et al.*, 2008 were probably the first to propose a framework for the design of GNNs. Then, Defferrard *et al.*, 2016 elaborated an efficient implementation of GNN using graph Fourier transform, and Kipf and Welling, 2017 have developed GNN in the semi-supervised learning context. Several works made nice connections between Personalized PageRank and GNNs: Klicpera *et al.*, 2019, Bojchevski *et al.*, 2020, Chien *et al.*, 2020. Recently, many works have been published on this topic and an interested reader can find comprehensive reviews in (Wu *et al.*, 2020; Zhou *et al.*, 2020).

In addition to the analysis presented in Section 5.4, the methods of random matrix theory have also been applied to semi-supervised learning in (Mai and Couillet, 2018, 2021).

With the advance of high-performance computing and cloud computing, one needs to consider parallel computation approaches to graph-based semi-supervised learning. A few examples of such approaches are presented in (Avrachenkov *et al.*, 2016a; Ravi and Diao, 2016; Chen *et al.*, 2020).

DOI: 10.1561/9781638280514.ch6

Chapter 6

Community Detection in Temporal Networks

Previous chapters focus on the study of static interactions, represented by a binary number or a positive weight. Nevertheless, in many application domains, interactions vary over time. The longitudinal nature of such datasets calls for replacing classical graph-based models with temporal network models represented by tensors (Holme and Saramäki, 2012; Kivelä *et al.*, 2014). We note that taking into account the temporal aspects not carefully, e.g., by aggregating or smoothing the temporal data along the time axis, can lead to a loss of valuable information.

The problem of community detection in temporal networks has recently attracted a considerable amount of attention from the scientific community. While it led to many interesting results, it also led to an explosion of disparate terminologies and algorithms.

In this chapter, we will first unify existing models of temporal networks with communities into a single framework. Then, we will show that the existing works can be grouped into two large categories: models with fixed community memberships and models with time-varying community memberships. We will then study each of these cases separately.

6.1 A General Model of Temporal Networks with Communities

6.1.1 Membership and Interaction Structures

We consider a block model for temporal networks with n nodes, K blocks and T temporal snapshots. The observed data consists of a list of T adjacency matrices $(A^1, \ldots, A^T)$, where each matrix $A^t \in \{0,1\}^{n\times n}$ describes a snapshot of the network at a particular time instant. Furthermore, at time t the node set is partitioned into K latent communities, and we denote by Z_{it} the label of node i at time t.

The matrix $Z \in [K]^{n\times T}$ represents the *membership structure*. Each column $Z_{\cdot t} \in [K]^n$ consists of the community labels of the nodes at a given time t, while the row $Z_{i\cdot} \in [K]^T$ represents the membership pattern of node i (*i.e.*, the evolution of the community label for node i).

We assume that the node membership patterns are independent and distributed according to a probability distribution p over $[K]^T$. Therefore,

$$\mathbb{P}(Z) = \prod_{i=1}^{n} p\left(Z_{i\cdot}\right). \tag{6.1}$$

Conditionally on the block membership structure Z, we want to generate a random tensor $A = \left(A_{ij}^t\right) \in \{0,1\}^{n\times n\times T}$, indexed by node pairs $\{i,j\}$ and time t, and verifying $A_{ij}^t = A_{ji}^t$, such that the pattern interactions between node pairs are independent. We denote by $B_{k^{1:T},\ell^{1:T}}\left(x^{1:T}\right)$ the probability of observing an interaction pattern $x^{1:T} \in \{0,1\}^T$ between a pair of nodes with block patterns $k^{1:T} = (k_1, \ldots, k_T) \in [K]^T$ and $\ell^{1:T} = (\ell_1, \ldots, \ell_T) \in [K]^T$. The *interaction structure* B is thus a collection of probability measures $B = \left(B_{k^{1:T},\ell^{1:T}}\right)$. Conditionally on B and Z, the law of the random tensor A is defined as

$$\mathbb{P}(A \mid Z, B) = \prod_{1\le i<j\le n} B_{Z_{i\cdot},Z_{j\cdot}}\left(A_{ij}^{1:T}\right). \tag{6.2}$$

The model (6.1)–(6.2) is the most general expression of a block model with n nodes and K clusters for a temporal network with T snapshots. Since the size of the block membership structure is $K \times T$, there is $\frac{(KT)^2}{2}$ choices of probability measures $B_{k^{1:T},\ell^{1:T}}$. Keeping this full generality leads to an over-complicated model. The following section details some particular cases of interest.

6.1.2 Examples of Temporal Network Models

Static memberships, dynamic interactions The block membership structure Z is static if the columns of the matrix Z are equal. Equivalently, the community labelling of each node does not vary over time. In that case, we can simply denote the static community labelling by a vector $z \in [K]^n$. Furthermore, the block interaction structure B reduces to an *interaction kernel* $f = (f_{k\ell})_{k,\ell \in [K]}$ which is a collection of probability distributions on $\mathcal{S} = \{0,1\}^T$ such that $f_{k\ell} = f_{\ell k}$. This defines the probability distribution

$$\mathbb{P}(A \mid z) = \prod_{1 \le i < j \le n} f_{z_i z_j}(A_{ij}) \tag{6.3}$$

of a symmetric interaction tensor $A \in \mathcal{S}^{n \times n}$ representing a temporal block model with static community structure. The model is *homogeneous* if the interaction kernel takes the form

$$f_{k\ell} = \begin{cases} f_{\text{in}}, & \text{if } k = \ell, \\ f_{\text{out}}, & \text{otherwise.} \end{cases}$$

Here f_{in} represents the distribution of the interactions within a block while the interactions across blocks are distributed according to f_{out}.

Example 6.1. Let $x = (x_1, \ldots, x_T) \in \{0,1\}^T$. A temporal SBM with static membership structure has *temporally independent interactions*, if for all $k, \ell \in [K]$, we have $f_{k\ell}(x) = \prod_{t=1}^{T} \mu_{k\ell}(x_t)$, where $\mu_{k\ell}$ is a probability distribution over $\{0,1\}$. A temporal block model with static membership structure and temporally independent interactions corresponds to T independent observations of a binary SBM.

Example 6.2. Let $x = (x_1, \ldots, x_T) \in \{0,1\}^T$. A temporal SBM with static memberships has *Markov interactions*, if for all $k, \ell \in [K]$, we have $f_{k\ell} = \mu_{k\ell}(x_1) \prod_{t=2}^{T} P_{k\ell}(x_{t-1}, x_t)$, where $\mu_{k\ell}$ is a probability distribution over $\{0,1\}$ and $P_{k\ell}$ is a 2-by-2 stochastic matrix representing the probability of transitions between two consecutive snapshots. If $P_{k\ell}(a,b) = \mu_{k\ell}(b)$ for all $k, \ell \in [K]$ and $a, b \in \{0,1\}$, then we recover the model described in Example 6.1.

Temporally independent interactions The block interaction structure is temporally independent, if at any time step t the binary interaction between nodes i and j is re-sampled according to the community labelling of i and j at time t. The law of the random tensor A is then given by

$$\mathbb{P}(A \mid Z, B) = \prod_{1 \le i < j \le n} Q_{Z_{it}, Z_{jt}}\left(A_{ij}^t\right) \tag{6.4}$$

where $Q = (Q_{k\ell})_{k,\ell \in [K]}$ is a set of distributions on $\{0,1\}$.

Markov membership structure Let $z^{1:T} = (z_1, \ldots, z_T) \in [K]^T$ denote a membership pattern and recall that p is the distribution of the nodes community assignment (see equation (6.1)). The model has a *Markov membership structure*, if

$$p(z) = \alpha_{z_1} \prod_{t=2}^{T} \pi_{z_{t-1}, z_t},$$

where α is the initial probability distribution on $[K]$ and π is a K-by-K matrix of transitions probabilities.

Example 6.3. Let $1_K = (1, \ldots, 1)^T$ be the $K \times 1$ vector of all ones. The model with a Markov membership structure, defined by $\alpha = \frac{1}{K} 1_K$ and $\pi = rI_K + \frac{1-r}{K} 1_K^T 1_K$, corresponds to a model, where:

- at initial time $t = 1$, the community labelling of all nodes are chosen independently and uniformly at random;
- at time $t \geq 2$, with a probability r a given node i remains in the community it was at time $t - 1$, while with probability $1 - r$ the node is assigned to a new community chosen uniformly at random.

6.2 Networks with Static Community Memberships

6.2.1 Recovery Thresholds in SBM with Markov Interaction

The model is called *Markov Stochastic Block Model*, if the community memberships are static and the temporal interactions are Markovian (see also Example 6.2). In a homogeneous Markov SBM, the interaction kernels take the form

$$\begin{aligned} f_{\text{in}} &= \mu_{x_1} P_{x_1, x_2} \cdots P_{x_{T-1}, x_T}, \\ f_{\text{out}} &= \nu_{x_1} Q_{x_1, x_2} \cdots Q_{x_{T-1}, x_T}, \end{aligned} \tag{6.5}$$

where μ, ν are the initial probability distributions on $\{0, 1\}$ and P, Q are the transition probability matrices on $\{0, 1\}$.

In the sparse regime, the probability of observing a non-zero interaction between any particular pair of nodes is small, *i.e.*,

$$\max\{\mu_1, \nu_1, P_{01}, Q_{01}\} \leq \rho. \tag{6.6}$$

One particular case is to assume that for some constants $u, v, p_{01}, q_{01} \in (0, \infty)$, we have

$$\mu_1 = u\rho, \quad \nu_1 = v\rho, \quad P_{01} = p_{01}\rho, \quad Q_{01} = q_{01}\rho. \tag{6.7}$$

Under this assumption, the expected number of 1's in a f-distributed signal is $\mathbb{E}\sum_{t=1}^{T} X_t \leq \mu_1 + (T-1)P_{01} = O(\rho T)$. Hence, when $\rho T = o(1)$, the probability of observing an interaction in any particular node pair is small.

The following proposition, whose proof can be found in (Avrachenkov *et al.*, 2022), states recovery conditions for a sparse Markov SBM when $n \gg 1$ and $T \gg 1$. The notions of consistent and strongly consistent estimators were defined in Section 4.4.3.

Proposition 6.1. *Consider a homogeneous Markov SBM composed of $n \gg 1$ nodes, $K \asymp 1$ blocks, $T \gg 1$ snapshots, where f_{in} and f_{out} are Markov chain distributions defined by* (6.5) *and satisfying* (6.7) *with a sparsity parameter ρ such that $\rho T \ll 1$. Suppose that P_{11} and Q_{11} are constants, such that $(P_{11}, Q_{11}) \neq (1,1)$, and that $(p_{01}, P_{11}) \neq (q_{01}, Q_{11})$. Let*

$$\tilde{I} = \left(\sqrt{p_{01}} - \sqrt{q_{01}}\right)^2 + 2\sqrt{p_{01}q_{01}}H_{11}^2,$$

where $H_{11}^2 = 1 - \frac{\sqrt{(1-P_{11})(1-Q_{11})}}{1-\sqrt{P_{11}Q_{11}}}$ is the squared Hellinger divergence between two geometric distributions with parameters P_{11} and Q_{11}. Then:

(i) *a consistent estimator does not exist for $\rho T \lesssim \frac{1}{n}$ and does exist for $\rho T \gg \frac{1}{n}$;*

(ii) *a strongly consistent estimator does not exist for $\rho T \ll \frac{\log n}{n}$ and does exist for $\rho T \gg \frac{\log n}{n}$;*

(iii) *in a critical regime with $\rho T = (1+o(1))\tau \frac{\log n}{n}$ for some constant τ, a strongly consistent estimator does not exist for $\tau \tilde{I} < K$ and does exist for $\tau \tilde{I} > K$.*

The quantity $\rho T \tilde{I}$ corresponds to the main term in the Taylor expansion of the Rényi divergence between two Markov chain distributions f_{in} and f_{out}. We refer to (Avrachenkov *et al.*, 2022) for more details and proofs.

Remark 6.1. We recall from Example 4.1 that consistent recovery in binary SBM requires $\rho \gg n^{-1}$. In particular, Proposition 6.1 shows that consistent recovery is possible even in a very sparse regime, as long as the number of snapshots is large enough. For example, if $\rho = \frac{1}{n}$, then T has to be at least of the order $\omega(1)$ for consistent recovery to be possible.

Remark 6.2. The regime with signal strength $\rho T = (1 + o(1))\tau \frac{\log n}{n}$ is an interesting critical regime. Indeed, in this regime, the phase transition for strong consistency occurs at $\tau \left(\sqrt{p_{01}} - \sqrt{q_{01}}\right)^2 + 2\tau\sqrt{p_{01}q_{01}}H_{11}^2 > K$. By comparison, the interesting regime for strong consistency in static SBM is $\rho = (1 + o(1))\frac{\log n}{n}$ (see Example 4.2).

6.2.2 Online Likelihood-based Algorithms for Markov Dynamics

In this section, we derive an algorithm for clustering temporal networks with static community memberships. We consider situations in which data arrives snapshot per snapshot and an online estimate of the community memberships is updated at each time step.

Model parameters are known Given $A^{1:t} = \left(A^1, \cdots, A^t\right)$, we define a log-likelihood ratio matrix by

$$M_{ij}^t = \log \frac{f_{\mathrm{in}}\left(A_{ij}^{1:t}\right)}{f_{\mathrm{out}}\left(A_{ij}^{1:t}\right)}, \tag{6.8}$$

where f_{in} and f_{out} are the intra- and inter-block interaction probabilities. In particular, the log of the probability of observing a graph sequence $A^{1:t}$ given node labelling z equals

$$\log \mathbb{P}\left(A \mid z\right) = \frac{1}{2} \sum_i \sum_{j \neq i} M_{ij}^t 1(z_j = z_i) + \frac{1}{2} \sum_i \sum_{j \neq i} f_{\mathrm{out}}(A_{ij}^{1:t}).$$

Therefore, given an assignment $\hat{z}^{t-1}$ computed from the observation of the $t-1$ first snapshots, one can compute a new assignment $\hat{z}^t$ such that node i is assigned to any block k which maximises

$$L_{i,k}^t = \sum_{j \neq i} M_{ij}^t \delta_{\hat{z}_j^{t-1} k}. \tag{6.9}$$

This formula is interesting only if the computation of M^t can be easily done from M^{t-1}. This is in particular the case for the Markovian evolution. Indeed, if f_{in} and f_{out} are given by (6.5), then the cumulative log-likelihood matrices defined in equation (6.8) can be computed recursively by $M^t = M^{t-1} + \Delta^t$, where

$$M_{ij}^1 = \log \frac{\mu}{\nu}\left(A_{ij}^1\right) \quad \text{and} \quad \Delta_{ij}^t = \log \frac{P}{Q}\left(A_{ij}^{t-1}, A_{ij}^t\right).$$

We summarises this in Algorithm 15. Let us emphasise that this algorithm works in an online adaptive fashion.

The time complexity (worst case complexity) of Algorithm 15 is $O(Kn^2T)$ plus the time complexity of the initial clustering. The space complexity is $O(n^2)$.

Algorithm 15: Online clustering for homogeneous Markov dynamics when the block interaction parameters are known.

Input: Interaction tensor (A_{ij}^t); block interaction parameters μ, ν, P, Q; number of communities K; static graph clustering algorithm algo.
Output: Node labelling $\hat{z} = \left(\hat{z}_1, \ldots, \hat{z}_n\right) \in [n]^K$.

Initialize: compute $\hat{z} \leftarrow \mathsf{algo}(A^1)$, and $M_{ij} \leftarrow \log \frac{\mu}{\nu}\left(A_{ij}^1\right)$ for $i, j = 1, \ldots, n$.
for $t = 2, \ldots, T$ **do**
  compute $\Delta_{ij} \leftarrow \log \frac{P}{Q}\left(A_{ij}^{t-1}, A_{ij}^t\right)$ for $i, j = 1, \ldots, n$;
  update $M \leftarrow M + \Delta$.
  for $i = 1, \ldots, n$ **do**
    set $L_{ik} \leftarrow \sum_{j \neq i} M_{ij}\, \delta_{\hat{z}_j k}$ for $k = 1, \ldots, K$;
    set $\hat{z}_i \leftarrow \arg\max_{1 \leq k \leq K} L_{ik}$.
Return: $\hat{z}$.

In addition, we note that:

- since at each time step, Δ_{ij} can take only one of four values, these four different values of Δ_{ij} can be precomputed and stored to avoid computing $n^2 T$ logarithms;
- the n-by-K matrix (L_{ik}) can be computed as a matrix product $L = M^0 Z$, where M^0 is the matrix obtained by zeroing out the diagonal of M, and Z is the one-hot representation of $\hat{z}$ such that $Z_{ik} = 1$, if $\hat{z}_i = k$, and zero, otherwise;
- for sparse networks the time and space complexity (average complexity) can be reduced by a factor of d/n where d is the average node degree in a single snapshot, by neglecting the $0 \to 0$ transitions and only storing nonzero entries.

Extension when the parameters are unknown Algorithm 15 requires the *a priori* knowledge of the interaction parameters. This is often not the case in practice and one has to learn the parameters during the process of recovering communities. In this section, we adapt Algorithm 15 by estimating the parameters on the fly.

Let $n_{ab}(i,j)$ be the observed number of transitions $a \to b$ in the interaction pattern between nodes i and j, and let $n_a(i,j) = \sum_b n_{ab}(i,j)$. Let $P(i,j)$ be the 2-by-2 matrix transition probabilities for the evolution of the pattern interaction for a node pair (i,j). By the law of large numbers (for stationary and ergodic random processes), the empirical transition probabilities

$$\widehat{P}_{ab}(i,j) = \frac{n_{ab}(i,j)}{n_a(i,j)} \tag{6.10}$$

are with high probability close to $P(i,j)$ for $T \gg 1$.

An estimator of P is obtained by averaging those probabilities over the pairs of nodes predicted to belong to the same community. More precisely, after t observed snapshots ($t \geq 2$), given a predicted community assignment $\hat{z}^t$, we define for $a, b \in \{0, 1\}$,

$$\widehat{P}^t_{ab} = \frac{1}{\left|\left\{(i,j) \, : \, \hat{z}^t_i = \hat{z}^{(t)}_j\right\}\right|} \sum_{(i,j) \, : \, \hat{z}^t_i = \hat{z}^{(t)}_j} \frac{n^t_{ab}(i,j)}{n^t_a(i,j)}, \tag{6.11}$$

where

$$n^t_{ab}(i,j) = \sum_{t'=1}^{t-1} 1\left(A^{t'}_{ij} = a\right) 1\left(A^{t'+1}_{ij} = b\right)$$

is the number of $a \to b$ transitions in the interaction pattern between nodes i and j (with $a, b \in \{0, 1\}$) seen during the t first snapshots and

$$n^t_a(i,j) = \sum_{b=0}^{1} n^t_{ab}(i,j).$$

Similarly,

$$\widehat{Q}^{(t)}_{ab} = \frac{1}{\left|\left\{(i,j) \, : \, \hat{z}^t_i \neq \hat{z}^t_j\right\}\right|} \sum_{(i,j) \, : \, \hat{z}^t_i \neq \hat{z}^t_j} \frac{n^{(t)}_{ab}(i,j)}{n^{(t)}_a(i,j)}, \tag{6.12}$$

is an estimator of Q_{ab}. Moreover, the quantities $n^{(t)}_{a,b}(i,j)$ can be updated inductively. Indeed,

$$n^{t+1}_{ab}(i,j) = n^{(t)}_{ab}(i,j) + 1\left(A^t_{ij} = a\right) 1\left(A^{t+1}_{ij} = b\right). \tag{6.13}$$

Finally, the initial distribution can also be estimated by averaging:

$$\hat{\mu}^t = \frac{1}{\left|\left\{(i,j) \, : \, \hat{z}^{(t)}_i = \hat{z}^t_j\right\}\right|} \sum_{(i,j) \, : \, \hat{z}^{(t)}_i = \hat{z}^t_j} A^t_{ij}$$

and similarly for $\hat{\nu}^t$. This leads to Algorithm 16, for clustering Markov SBM when only the number of communities K is known. Note that to save computation time, we can choose not to update the parameters at each time step.

Algorithm 16: Online clustering for homogeneous Markov dynamics when the block interaction parameters are unknown.

Input: Observed graph sequence $X^{1:T} = \left(X^1, \ldots, X^T\right)$; number of communities K; static graph clustering algorithm algo.

Output: Node labelling $\hat{z} = \left(\hat{z}_1, \ldots, \hat{z}_n\right)$.

Initialize:

- Compute $\hat{z} \leftarrow \mathsf{algo}\left(X^1\right)$;
- Let $n_{ab}(i,j) \leftarrow 0$ for $i, j \in [N]$ and $a, b \in \{0, 1\}$.

Update:

for $t = 2, \cdots, T$ **do**

  For every node pair (ij), update $n_{ab}(i,j)$ using (6.13);

  Compute $\widehat{P}, \widehat{Q}$ using (6.11) and (6.12);

  Compute M such that $M_{ij} = \sum_{a,b} n_{ab}(i,j) \log \frac{\widehat{P}_{ab}}{\widehat{Q}_{ab}}$.

  for $i = 1, \cdots, n$ **do**

    Set $L_{i,k} \leftarrow \sum_{j \neq i} M_{ij} 1\left(\hat{z}_j = k\right)$ for all $k = 1, \ldots, K$

    Set $\hat{z}_i \leftarrow \arg\max_{1 \leq k \leq K} L_{i,k}$

Numerical results

Evolution of accuracy with the number of snapshots Let us first study numerically the effect of the initialization step. We plot in Figure 6.1 the evolution of the averaged accuracy obtained when we run Algorithm 15 on 50 realizations of a Markov SBM, where the initialization is done either using spectral clustering or by random guessing. Obviously, when spectral clustering works well (see Figure 6.1(a)), it is preferable to use it rather than a random guess. Nonetheless, it is striking to see that when the initial spectral clustering gives a bad accuracy, then the likelihood method can overcome it. For example, in Figure 6.1(b), the initial clustering with spectral clustering on the first snapshot is really bad (accuracy $\approx 50\%$, hence not much better than random guessing), Algorithm 15 does overcome this and reaches perfect clustering after a few snapshots. In that particular setting, there is no advantage in using spectral clustering instead of random guessing. Additionally, random guessing is faster than spectral clustering.

Unknown interaction parameters We next show in Figure 6.2 the comparison of accuracy obtained by Algorithm 15 (with known interaction parameters) and by Algorithm 16 (with unknown interaction parameters). We note that in all

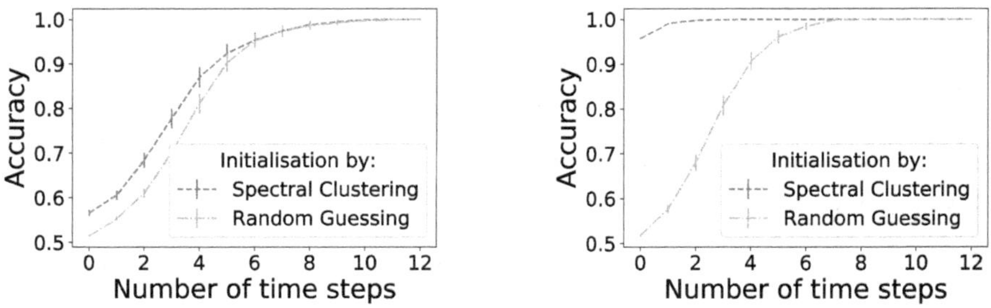

(a) $\mu_1 = 2.5\,\frac{\log n}{n}$ ($T^*_{\text{theo}} = 14$)

(b) $\mu_1 = 4.0\,\frac{\log n}{n}$ ($T^*_{\text{theo}} = 11$)

Figure 6.1. Evolution of the averaged accuracy given by Algorithm 15 when the initialisation is done via spectral clustering or random guessing. The synthetic graphs are Markov SBM with $n = 500$ nodes (equally divided into two clusters), and with parameters $\nu_1 = 1.5\,\frac{\log n}{n}$, $P_{11} = 0.7$ and $Q_{11} = 0.3$. Accuracy is averaged over 50 realisations, and the error bars represent the standard error. T^*_{theo} is the theoretical minimum number of time steps needed to get above the strong consistency threshold.

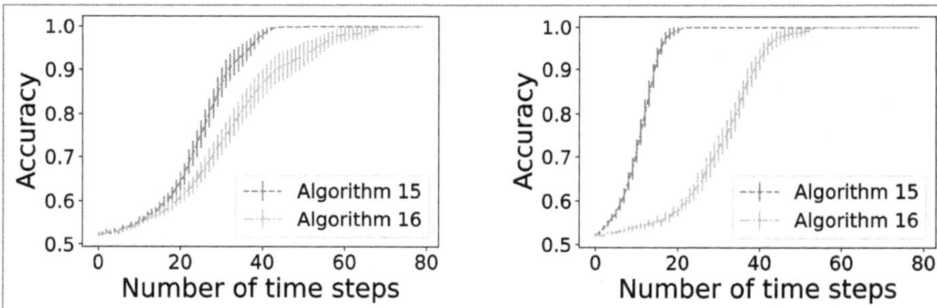

(a) $\mu_1 = 0.004$.

(b) $\nu_1 = 0.016$.

Figure 6.2. Comparison of accuracy obtained by online Algorithms 15 and 16 on Markov SBMs with $N = 400$, $K = 2$ and $\nu_1 = 0.004$. Results are averaged over 25 Markov SBMs and error bars show the standard errors.

the following numerical experiments, we will chose sparse settings in which spectral clustering on a single snapshot do not provide more information than a blind random guess. While Algorithm 15 provides better performance (as expected as it does not have to estimate the Markov chain transition probabilities), Algorithm 16 also provides an excellent accuracy using more snapshots.

Let us finally study the performance of Algorithm 16 in Markov SBMs for which the expected degree in a given temporal layer is less than 1. This corresponds to an extremely sparse regime. Nontheless, as we see in Figure 6.3, Algorithm 15 performs well, even when $\mu_1 = \nu_1$, as long as $P_{11} \neq Q_{11}$ (see Figure 6.3(a)). This shows that Algorithm 16 recovers the communities very well, even in the most challenging regimes.

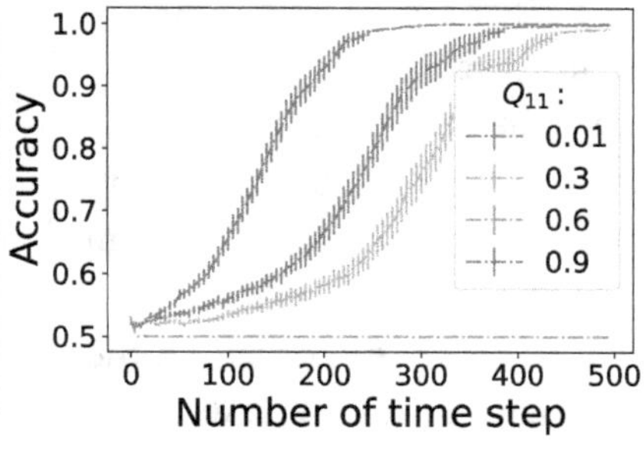

(a) $\mu_1 = \frac{0.1}{N}$.

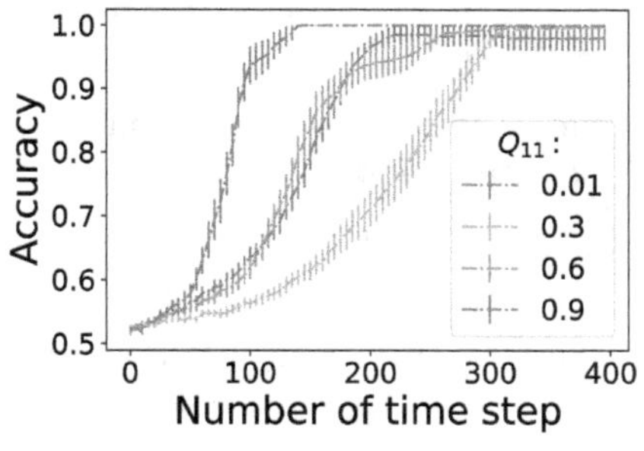

(b) $\mu_1 = \frac{0.3}{N}$

Figure 6.3. Evolution of the accuracy with the number of snapshots obtained by Algorithm 16 in an extremely sparse setting. We draw Markov SBM with $N = 300$ nodes, two communities of same size and parameters $\nu_1 = \frac{0.1}{N}$ and $P_{11} = 0.6$. The different curves show the average on 25 Markov SBMs and the errors bars correspond to the empirical standard errors.

6.2.3 Spectral Methods for Clustering Temporal Networks

We introduced in Section 4.1 spectral methods for static graphs as relaxations of various combinatorial minimisation problems. The simplest of those problems is the min Cut, *i.e.*,

$$\arg\min_{z \in [K]^n} \mathrm{Cut}(A, z),$$

where the arg min runs over all possible node labellings $z \in [K]^n$ of the node set $[n]$, and where

$$\mathrm{Cut}(A, z) = \sum_{i<j:\ z_i \neq z_j} A_{ij}.$$

Let us now consider a temporal network represented by its list of adjacency matrices $\left(A^1, \ldots, A^T\right)$. If we assume that the temporal snapshots A^t are independent of each others, one could simply generalize the classical min Cut problem by considering

$$\arg\min_{z \in [K]^n} \sum_{t=1}^{T} \mathrm{Cut}\left(A^t, z\right).$$

Since $\sum_{t=1}^{T} \mathrm{Cut}\left(A^t, z\right) = \mathrm{Cut}\left(\sum_{t=1}^{T} A^t, z\right)$, we would then apply a spectral method on the time-aggregated graph (that is, the weighted graph represented by the adjacency matrix $\sum_{t=1}^{T} A^t$).

Unfortunately, this fails at taking into account the time-correlation in the interaction patterns between nodes. As an example, consider a network in which the inter-community interactions are sparse and temporally independent (hence forming spikes), while the intra-community interactions are strongly correlated in time. Consider two node pairs whose interaction patterns are given by $x_1 = (0,1,0,0,0,0,1,0,0,1)$ and $x_2 = (0,0,1,1,1,0,0,0,0,0)$. Since $\|x_1\|_1 = \|x_2\|_1 = 3$, we see that simple time-aggregation is agnostic to the different time-patterns between time series x_1 and x_2 and the important information is lost.

A possible correction is to account for the *persistent links*. Indeed, in the above example, since x_1 (resp., x_2) has zero (resp., two) transitions $1 \to 1$, we might guess than x_2 comes from an interaction between nodes belonging to same community. Formally, this can be done by considering

$$\arg\min_z \sum_{t=1}^{T} \operatorname{Cut}\left(A^t, z\right) + \alpha \sum_{t=2}^{T} \operatorname{PerCut}\left(A^{t-1}, A^t, z\right),$$

where

$$\operatorname{PerCut}\left(A^{t-1}, A^t, z\right) = \sum_{i,j\,:\, z_i \neq z_j} A_{ij}^{t-1} A_{ij}^t$$

counts the number of persistent links in the cut from time $t-1$ to time t. We further notice that

$$\operatorname{PerCut}\left(A^{t-1}, A^t, z\right) = \operatorname{Cut}\left(A^{t-1} \odot A^t, z\right)$$

where $\odot$ denotes the matrix element-wise product. The following section justifies the intuition of considering persistent edges for clustering temporal networks.

Degree-corrected temporal SBM with Markov edge dynamics Let us firstly present a degree-corrected version of the Markov SBM. A degree-corrected temporal stochastic block model with n nodes, K blocks and T snapshots can be described by the probability distribution

$$\mathbb{P}(A \mid Z, F, \theta) = \prod_{1 \leq i < j \leq n} F_{z_i z_j}^{\theta_i \theta_j}\left(A_{ij}^1, \ldots, A_{ij}^T\right) \tag{6.14}$$

of a symmetric adjacency tensor $A \in \{0,1\}^{n \times n \times T}$ with zero diagonal entries, where $z = (z_1, \ldots, z_n)$ is a community assignment with $z_i \in [K]$ indicating the community of node i, $F = \left(F_{k\ell}^{xy}\right)$ is a collection of probability distributions over $\{0,1\}^T$, and $\theta = (\theta_1, \ldots, \theta_n)$ is a vector of node-specific degree correction parameters, with $0 \leq \theta_i < \infty$.

In the following, we will restrict ourselves to homogeneous inter-block interactions with Markov edge dynamics, for which the nodes' static community labellings are sampled uniformly at random from the set $[K]$ of all node labellings, and

$$F_{z_i z_j}^{\theta_i\theta_j}(x) = \begin{cases} \mu_{x_1}^{\theta_i\theta_j} \prod_{t=2}^{T} P_{x_{t-1},x_t}^{\theta_i\theta_j} & \text{if } z_i = z_j, \\ \nu_{x_1}^{\theta_i\theta_j} \prod_{t=2}^{T} Q_{x_{t-1},x_t}^{\theta_i\theta_j} & \text{otherwise}, \end{cases} \tag{6.15}$$

with initial distributions

$$\mu^{\theta_i\theta_j} = \begin{pmatrix} 1 - \theta_i\theta_j\mu_1 \\ \theta_i\theta_j\mu_1 \end{pmatrix}, \quad \nu^{\theta_i\theta_j} = \begin{pmatrix} 1 - \theta_i\theta_j\nu_1 \\ \theta_i\theta_j\nu_1 \end{pmatrix},$$

and transition probability matrices

$$P^{\theta_i\theta_j} = \begin{pmatrix} 1 - \theta_i\theta_j P_{01} & \theta_i\theta_j P_{01} \\ 1 - P_{11} & P_{11} \end{pmatrix}, \quad Q^{\theta_i\theta_j} = \begin{pmatrix} 1 - \theta_i\theta_j Q_{01} & \theta_i\theta_j Q_{01} \\ 1 - Q_{11} & Q_{11} \end{pmatrix}.$$

The parameters $\theta_i, i = 1, \ldots, n$ account for the fact that some nodes can be more prone than others to start new connections, similarly to the degree-corrected block model (Karrer and Newman, 2011). To keep the model simple, we do not add degree correction parameters in front of P_{11}; hence once a connection started, the probability to keep it active is simply P_{11} or Q_{11}. Moreover, we assume that $\min_{i,j}\{\theta_i\theta_j\delta\} \leq 1$, where $\delta = \max\{\mu_1, \nu_1, P_{01}, Q_{01}\}$. Finally, we normalise the degree correction parameters so that $\sum_i 1(z_i = k)\theta_i = \sum_i 1(z_i = k)$ for all k.

Maximum likelihood estimator

Proposition 6.2 (Avrachenkov *et al.*, 2021b). *Let* $\rho_a^{\theta_i\theta_j} = \log \frac{\mu_a^{\theta_i\theta_j}}{\nu_a^{\theta_i\theta_j}}$ *and* $\ell_{ab}^{\theta_i\theta_j} = \log \frac{P_{ab}^{\theta_i\theta_j}}{Q_{ab}^{\theta_i\theta_j}} - \log \frac{P_{00}^{\theta_i\theta_j}}{Q_{00}^{\theta_i\theta_j}}$. *A maximum likelihood estimator for the Degree Corrected Markov SBM defined by* (6.14)–(6.15) *is any community assignment* $\hat{z}$ *that maximises*

$$\sum_{\substack{i,j \\ z_i=z_j}} \left\{ A_{ij}^1 \left(\rho_1^{\theta_i\theta_j} - \rho_0^{\theta_i\theta_j} \right) + \rho_0^{\theta_i\theta_j} + \left(A_{ij}^1 - A_{ij}^T \right) \ell_{10}^{\theta_i\theta_j} \right\} + \sum_{\substack{i,j \\ z_i=z_j}} \sum_{t=2}^{T}$$

$$\left\{ \left(\ell_{01}^{\theta_i\theta_j} + \ell_{10}^{\theta_i\theta_j} \right) \left(A_{ij}^t - A_{ij}^{t-1} A_{ij}^t \right) + \ell_{11}^{\theta_i\theta_j} A_{ij}^{t-1} A_{ij}^t - \log \frac{Q_{00}^{\theta_i\theta_j}}{P_{00}^{\theta_i\theta_j}} \right\}$$

over all community assignments $z \in [K]^n$.

Proof. By the temporal Markov property, the log-likelihood of the model can be written as $\log \mathbb{P}(A \mid z, \theta) = \log \mathbb{P}(A^1 \mid z, \theta) + \sum_{t=2}^{T} \mathbb{P}(A^t \mid A^{t-1}, z, \theta)$. By denoting $\rho_a^{\theta_i\theta_j} = \log \frac{\mu_a^{\theta_i\theta_j}}{\nu_a^{\theta_i\theta_j}}$, we find that

$$\begin{aligned}\log \mathbb{P}(A^1 \mid z, \theta) &= \frac{1}{2} \sum_{i,j} \sum_{a} \delta(A_{ij}^1, a) \Big(\delta(z_i, z_j) \rho_a^{\theta_i\theta_j} + \log \nu_a^{\theta_i\theta_j} \Big) \\ &= \frac{1}{2} \sum_{i,j} \delta(z_i, z_j) \sum_{a} \delta(A_{ij}^1, a) \rho_a^{\theta_i\theta_j} + c_1(A),\end{aligned}$$

where $c_1(A) = \frac{1}{2} \sum_{i,j} \sum_a \delta(A_{ij}^1, a) \log \nu_a^{\theta_i\theta_j}$ does not depend on the community structure. Similarly, by denoting $R_{ab}^{\theta_i\theta_j} = \log \frac{P_{ab}^{\theta_i\theta_j}}{Q_{ab}^{\theta_i\theta_j}}$, we find that $\log \mathbb{P}(A^t \mid A^{t-1}, z, \theta)$ is equal to

$$\begin{aligned}&\frac{1}{2} \sum_{i,j} \sum_{a,b} \delta(A_{ij}^{t-1}, a) \delta(A_{ij}^t, b) \Big(\delta(z_i, z_j) R_{ab}^{\theta_i\theta_j} + \log Q_{ab}^{\theta_i\theta_j} \Big) \\ &\quad = \frac{1}{2} \sum_{i,j} \delta(z_i, z_j) \sum_{a,b} \delta(A_{ij}^{t-1}, a) \delta(A_{ij}^t, b) R_{ab}^{\theta_i\theta_j} + c_t(A),\end{aligned}$$

where $c_t(A) = \frac{1}{2} \sum_{i,j} \sum_{a,b} \delta(A_{ij}^{t-1}, a) \delta(A_{ij}^t, b) \log Q_{ab}^{\theta_i\theta_j}$ does not depend on the community structure. Simple calculations show that

$$\sum_a \delta(A_{ij}^1, a) \rho_a^{\theta_i\theta_j} = A_{ij}^1 (\rho_1^{\theta_i\theta_j} - \rho_0^{\theta_i\theta_j}) + \rho_0^{\theta_i\theta_j},$$

and that $\sum_{a,b} \delta(A_{ij}^{t-1}, a) \delta(A_{ij}^t, b) R_{ab}^{\theta_i\theta_j}$ is equal to

$$\begin{aligned}&R_{00}^{\theta_i\theta_j} + A_{ij}^{t-1} (R_{10}^{\theta_i\theta_j} - R_{00}^{\theta_i\theta_j}) + A_{ij}^t (R_{01}^{\theta_i\theta_j} - R_{00}^{\theta_i\theta_j}) \\ &\qquad + A_{ij}^{t-1} A_{ij}^t (R_{11}^{\theta_i\theta_j} - R_{01}^{\theta_i\theta_j} - R_{10}^{\theta_i\theta_j} + R_{00}^{\theta_i\theta_j}) \\ &R_{00}^{\theta_i\theta_j} + A_{ij}^{t-1} \ell_{10}^{\theta_i\theta_j} + A_{ij}^t \ell_{01}^{\theta_i\theta_j} \\ &\qquad + A_{ij}^{t-1} A_{ij}^t (\ell_{11}^{\theta_i\theta_j} - \ell_{01}^{\theta_i\theta_j} - \ell_{10}^{\theta_i\theta_j}).\end{aligned}$$

By collecting the above observations, we find that $\log \mathbb{P}(A \mid z, \theta)$ is equal to

$$c(A) + \frac{1}{2} \sum_{\substack{i,j \\ z_i = z_j}} \left\{ A_{ij}^1 (\rho_1^{\theta_i \theta_j} - \rho_0^{\theta_i \theta_j}) + \rho_0^{\theta_i \theta_j} + (A_{ij}^1 - A_{ij}^T) \ell_{10}^{\theta_i \theta_j} \right\}$$

$$+ \frac{1}{2} \sum_{\substack{i,j \\ z_i = z_j}} \sum_{t=2}^{T}$$

$$\left\{ (\ell_{01}^{\theta_i \theta_j} + \ell_{10}^{\theta_i \theta_j}) \left(A_{ij}^t - A_{ij}^{t-1} A_{ij}^t \right) + \ell_{11}^{\theta_i \theta_j} A_{ij}^{t-1} A_{ij}^t - \log \frac{Q_{00}^{\theta_i \theta_j}}{P_{00}^{\theta_i \theta_j}} \right\},$$

where $c(A) = \sum_t c_t(A)$ does not depend on z. Hence the claim follows. □

The MLE derived in Proposition 6.2 is more complex than that obtained by summing all snapshots independently. In particular, the terms $A_{ij}^{t-1} A_{ij}^t$ account for *persistent edges* over two consecutive snapshots. Denote by $A_{\text{pers}}^t = A^{t-1} \odot A^t$ the entrywise product of adjacency matrices A^{t-1} and A^t. Then A_{pers}^t is the adjacency matrix of the graph containing the persistent edges between $t-1$ and t, and $A_{\text{new}}^t = A^t - A_{\text{pers}}^t$ corresponds to the graph containing the *freshly appearing edges* between time $t-1$ and time t.

Assuming that the number of snapshots T is large, we can ignore the boundary terms, and the MLE expressed in Proposition 6.2 reduces to maximising

$$\sum_{t=2}^{T} \sum_{\substack{i,j \\ z_i = z_j}} \left(\left(\ell_{01}^{\theta_i \theta_j} + \ell_{10}^{\theta_i \theta_j} \right) \left(A_{ij}^t - A_{ij}^{t-1} A_{ij}^t \right) + \ell_{11}^{\theta_i \theta_j} A_{ij}^{t-1} A_{ij}^t - \log \frac{Q_{00}^{\theta_i \theta_j}}{P_{00}^{\theta_i \theta_j}} \right).$$

This expression can be further simplify to be expressed as a regularized modularity. Recall given a weighted graph W, a partition z and a resolution parameter γ, the regularised modularity is defined as (see Section 4.2 and equation (4.21))

$$\mathcal{M}(W, z, \gamma) = \sum_{i,j} \delta(z_i, z_j) \left(W_{ij} - \gamma \frac{d_i d_j}{2m} \right),$$

where $d_i = \sum_j W_{ij}$ and $m = \frac{1}{2} \sum_i d_i$.

Lemma 6.1. *Suppose that $P^{\theta_i \theta_j}$ and $Q^{\theta_i \theta_j}$ are non-degenerate, and $\mu^{\theta_i \theta_j}$ (resp., $\nu^{\theta_i \theta_j}$) is the stationary distribution of $P^{\theta_i \theta_j}$ (resp., $Q^{\theta_i \theta_j}$). In a sparse setting, where P_{01} and Q_{01} are small, the MLE approximately maximises $\mathcal{M}(W, z, \gamma)$, where W is defined* *by*

$$W = \sum_{t=2}^{T} \left(\alpha A^t_{\text{new}} + \beta A^t_{\text{pers}} \right), \tag{6.16}$$

with

$$\alpha = \log \frac{P_{01}}{Q_{01}} + \log \frac{1 - P_{11}}{1 - Q_{11}}, \quad \textit{and} \quad \beta = \log \frac{P_{11}}{Q_{11}}, \tag{6.17}$$

and $\gamma = (P_{01} - Q_{01}) \frac{\alpha(\mu_1 + (K-1)\nu_1) + (\beta - \alpha)(\mu_1 P_{11} + (K-1)\nu_1 Q_{11})}{K}$.

Proof. Because $P_{01}, Q_{01} = o(1)$, a first-order Taylor expansion yields

$$\log \frac{1 - \theta_i\theta_j Q_{01}}{1 - \theta_i\theta_j P_{01}} = \theta_i\theta_j \left(P_{01} - Q_{01}\right) + o\left(P_{01}^2 + Q_{01}^2\right),$$

as well as $\ell_{01}^{\theta_i\theta_j} \approx \log \frac{P_{01}}{Q_{01}}$, $\ell_{10}^{\theta_i\theta_j} \approx \log \frac{1-P_{11}}{1-Q_{11}}$ and $\ell_{11}^{\theta_i\theta_j} \approx \log \frac{P_{11}}{Q_{11}}$. Using these approximations in the MLE expression leads to maximising

$$\sum_{t=2}^{T} \sum_{i,j} \delta(z_i, z_j) \left(\tilde{a}^t_{ij} - \theta_i\theta_j \left(P_{01} - Q_{01}\right) \right) \tag{6.18}$$

where $\tilde{a}^t_{ij} = \alpha \left(A^t_{\text{new}}\right)_{ij} + \beta \left(A^t_{\text{pers}}\right)_{ij}$. Since μ and ν are stationary distributions,

$$\mathbb{E}\left(A^t_{\text{new}}\right)_{ij} = \begin{cases} \theta_i\theta_j\mu_1(1 - P_{11}) & \text{if } z_i = z_j, \\ \theta_i\theta_j\nu_1(1 - Q_{11}) & \text{otherwise,} \end{cases}$$

$$\mathbb{E}\left(A^t_{\text{pers}}\right)_{ij} = \begin{cases} \theta_i\theta_j\mu_1 P_{11} & \text{if } z_i = z_j, \\ \theta_i\theta_j\nu_1 Q_{11} & \text{otherwise.} \end{cases}$$

Therefore, using $W_{ij} = \sum_{t=2}^{T} \tilde{a}_{ij}$, we obrain

$$\mathbb{E} W_{ij} = \begin{cases} (T-1)\theta_i\theta_j\mu_1 \left(\alpha(1 - P_{11}) + \beta P_{11}\right) & \text{if } z_i = z_j, \\ (T-1)\theta_i\theta_j\nu_1 \left(\alpha(1 - Q_{11}) + \beta Q_{11}\right) & \text{otherwise.} \end{cases}$$

Since the community labelling is sampled uniformly at random, and θ_i's are properly normalised, the expected degree $\bar{d}_i$ is equal to

$$(T-1)\theta_i n \frac{\mu_1 \left(\alpha(1 - P_{11}) + \beta P_{11}\right) + (K-1)\nu_1 \left(\alpha(1 - Q_{11}) + \beta Q_{11}\right)}{K},$$

together with $\bar{m} = \frac{n^2}{2}\frac{\mu_1(\alpha(1-P_{11})+\beta P_{11})+(K-1)\nu_1(\alpha(1-Q_{11})+\beta Q_{11})}{K}$. Hence, we observe that $\theta_i\theta_j(P_{01} - Q_{01}) = \gamma\frac{\bar{d}_i\bar{d}_j}{2\bar{m}}$ where $\gamma = (P_{01} - Q_{01})(T-1)\frac{\mu_1(\alpha(1-P_{11})+\beta P_{11})+(K-1)\nu_1(\alpha(1-Q_{11})+\beta Q_{11})}{K}$. We end the proof using equation (6.18). □

Temporal spectral clustering combining new and persistent edges Following our analysis of the previous section, the MLE is approximately given by the solution of

$$\arg\max_{z\in[K]^n} \mathcal{M}(W, z, \gamma),$$

where W is defined in Equation (6.16) and γ is an appropriate resolution parameter. This optimisation problem is NP-complete in general (Brandes *et al.*, 2007), but can be approximately solved by continuous relaxation. We can choose the relaxation so that the optimisation problem reduces to normalised spectral clustering algorithm on the weighted graph W (see Section 4.4.2). We note that in order to compute the normalized Laplacian of W, we should restrict $\alpha, \beta \geq 0$, which is not necessarily guaranteed by formula (6.17). We summarize this in Algorithm 17.

Algorithm 17: Spectral clustering for temporal networks with Markov edge dynamics and static node labelling.

Input: adjacency matrices $A^1, \cdots, A^T$, number of clusters K, parameters α, β.

Output: predicted community labels $\hat{z} \in [K]^N$.

Process:

- let $W = \sum_{t=2}^{T}(\alpha A^t_{\text{new}} + \beta A^t_{\text{pers}})$, where $A^t_{\text{new}} = A^t - A^{t-1} \odot A^t$ and $A^t_{\text{pers}} = A^{t-1} \odot A^t$;
- compute $\mathcal{L} = I_n - D^{-1/2}WD^{-1/2}$ where $D = \text{diag}(W1_n)$;
- compute $\widehat{X} \in \mathbb{R}^{N\times K}$ whose columns consist of the K orthonormal eigenvectors of $\mathcal{L}$ associated to the K smallest eigenvalues.

Return: $\hat{z} \leftarrow$ k-means $\left(D^{-1/2}\widehat{X}, K\right)$.

Numerical results

Synthetic data We first examine the effect of the choice of the parameters α and β in Algorithm 17. For this, we let $\alpha = 1$ and we plot in Figure 6.4 the averaged accuracy obtained on 25 realizations of stochastic block models with Markov edge dynamics for various β. While spectral clustering on the time-aggregated graph

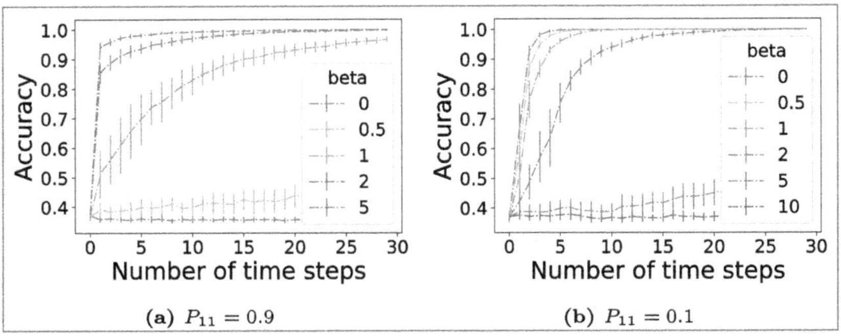

Figure 6.4. Accuracy of Algorithm 17 on a temporal SBM with 300 nodes, $K = 3$ blocks, and a stationary Markov edge evolution $\mu_1 = 0.04$, $\nu_1 = 0.02$ and $Q_{11} = 0.3$. The results are averaged over 25 synthetic graph realizations, and error bars show the standard deviation.

(corresponding to $\beta = 1$) works well, it is striking to notice that other values of β give even better results. The choice of β depends on the probabilities of persistent interactions. For example, if $P_{11} > Q_{11}$ (Figure 6.4(a)), then $\beta > 1$ is preferred, while if $P_{11} < Q_{11}$ (Figure 6.4(b)), the choices of large β are penalized. This is in accordance with the recommended values of α and β derived in formula (6.17).

Social networks of high school students We investigate three data sets collected during three consecutive years from a high school Lycée Thiers in Marseilles, France (Fournet and Barrat, 2014; Mastrandrea *et al.*, 2015). We presented these data sets in the introduction. In particular, nodes correspond to students, interactions to close-proximity encounters, and communities to classes, with dimensions given in Table 1.1.

We make a hypothesis that the temporal characteristics of the interactions are similar each year. We then use the 2011 data set to estimate the transition probability matrices P and Q, and use these for clustering the 2012 and 2013 data sets. We assume that $\theta_i = 1$ (no degree correction). A standard estimator of Markov chain transition probability matrices (Billingsley, 1961) gives

$$\widehat{P} = \begin{pmatrix} 0.9992 & 0.0008 \\ 0.37 & 0.63 \end{pmatrix} \quad \text{and} \quad \widehat{Q} = \begin{pmatrix} 0.999967 & 3.3 \times 10^{-5} \\ 0.48 & 0.52 \end{pmatrix}.$$

Using (6.17), leads to $\hat{\alpha} = 2.9$ and $\hat{\beta} = 0.18$. We observe in Figure 6.5(b) that this choice of parameters gives a better accuracy on the 2013 data set than simply applying spectral clustering on the time-aggregated graph ($\alpha = \beta = 1$). For the 2012 data set (Figure 6.5(a)), this improvement is not so clearly visible.

To understand why Algorithm 17 performs better for 2013 than for 2012, we have listed in Table 6.1 temporal transition probabilities and clustering weights $\widehat{\alpha}, \widehat{\beta}$ estimated separately for each data set. For year 2012, the difference between

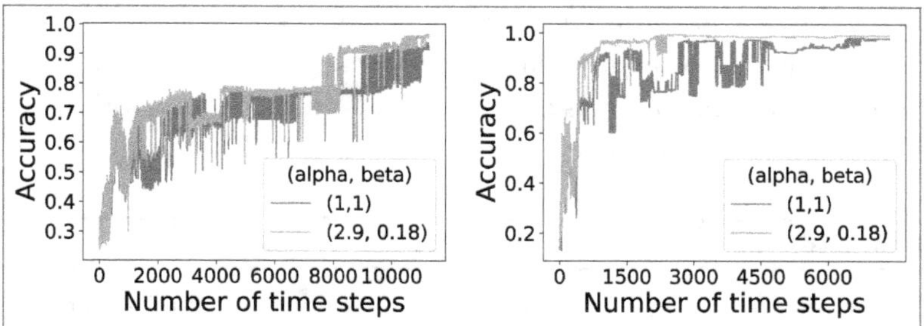

(a) Year 2012 (b) Year 2013

Figure 6.5. Accuracy of Algorithm 17 on the 2012 and 2013 high school data sets, using uniform $\alpha = \beta = 1$ (blue) and adjusted α, β, whose values are predicted using 2011 data (orange).

Table 6.1. Markov chain transition probabilities and adjusted clustering weights estimated separately for each data set.

Dataset	$\widehat{P}_{01}$	$\widehat{Q}_{01}$	$\widehat{P}_{11}$	$\widehat{Q}_{11}$	$\widehat{\alpha}$	$\widehat{\beta}$	$\widehat{\beta}/\widehat{\alpha}$
2011	0.00080	0.000033	0.63	0.52	2.9	0.58	0.060
2012	0.00050	0.000011	0.57	0.56	3.8	0.01	0.003
2013	0.00150	0.000014	0.64	0.40	4.5	0.07	0.015

intra-community edge persistence $\widehat{P}_{11}$ and inter-community edge persistence $\widehat{Q}_{11}$ is small, implying that persistent edges do not add much extra information for distinguishing communities ($\widehat{\beta} \approx 0$). For years 2011 and 2013, this difference is larger, manifesting that edge persistence contains information that can be employed to recover communities with a higher accuracy.

6.2.4 Clustering for Long Time Horizon Using Empirical Transition Rates

We continue to study the temporal SBM with static memberships and homogeneous Markov interaction kernels, as defined in Example 6.2. We denote by P, Q the transition probability matrices. Let us consider the situation when the number of snapshots T goes to infinity while N remains bounded. The main idea is to use the ergodicity of the Markov chains to estimate the parameters using standard techniques, and then perform inference. For now, we will assume that the interaction parameters P, Q are known, but K is unknown. We refer to Remark 6.3 when P, Q are unknown as well.

Recall that formula (6.10) gave consistent estimators for $P(i,j)$, the matrix of transition probabilities for the evolution of the pattern interaction between a node pair (i,j). Then, once all $P(i,j)$ are known with a good precision, we can use our knowledge of P, Q to distinguish whether nodes i and j are in the same block or not, and use this data to construct a similarity graph on the set of nodes. This leads to Algorithm 18 which does not require *a priori* knowledge about the number of blocks, but instead estimates it as a by-product. Note that this algorithm is tailor-made for homogeneous interaction arrays.

Algorithm 18: Clustering by empirical transition rates.

Input: observed interaction tensor $\left(A_{ij}^t\right)$; transition probability matrices P, Q.

.**Output**: estimated node labelling $\hat{z} = (\hat{z}_1, \ldots, \hat{z}_n)$; estimated number of communities $\widehat{K}$.

$V \leftarrow \{1, \ldots, n\}$ and $E \leftarrow \emptyset$.

for *all unordered node pairs ij* **do**

  compute $\widehat{P}_{ab}(i,j)$ for $a, b = 0, 1$ using (6.10).

  if $|\widehat{P}_{ab}(i,j) - P_{ab}| \leq \frac{1}{2} |P_{ab} - Q_{ab}|$ *for some* a, b **then**

    set $E \leftarrow E \cup \{ij\}$.

Compute $\mathcal{C} \leftarrow$ set of connected components in $G = (V, E)$ and set $\widehat{K} \leftarrow |\mathcal{C}|$ and $(C_1, \ldots, C_{\widehat{K}}) \leftarrow$ members of $\mathcal{C}$ listed in arbitrary order.

for $i = 1, \ldots, n$ **do**

  $\hat{z}_i \leftarrow$ unique k for which $C_k \ni i$.

Theorem 6.2. *Consider a homogeneous Markov SBM with n nodes, K communities and T snapshots. Assume that n is fixed, and the transition probability matrices P, Q are known. Then with high probability Algorithm 18 correctly classify every node when T goes to infinity, as long as the evolution is not static and $P \neq Q$.*

Proof. For $a, b \in \{0, 1\}$, let $n_a(i,j) = \sum_b n_{ab}(i,j)$ where $n_{ab}(i,j)$ counts the observed number of transitions $a \to b$ between a node pair (i,j). The distribution of the random variable $\xi_{ab}(i,j) = \frac{n_{ab}(i,j) - n_a(i,j)P_{ab}(i,j)}{\sqrt{n_a(i,j)}}$ tends to a normal distribution with the zero mean and finite variance given by $\lambda_{(ab),(cd)} = \delta_{ac}\big(\delta_{bd}P_{ab}(i,j) - P_{ab}(i,j)P_{a,d}(i,j)\big)$ (see Billingsley, 1961, Theorem 3.1 and formula (3.13)). Therefore, for any $\alpha > 0$,

$$\mathbb{P}\Big(|\widehat{P}_{ab}(i,j) - P_{ab}(i,j)| \geq \alpha\Big) = \mathbb{P}\Big(|\xi_{ab}(i,j)| \geq \alpha\sqrt{n_a(i,j)}\Big), \tag{6.19}$$

and this quantity goes to zero as T goes to infinity.

From model identifiability, $P \neq Q$. Therefore, without loss of generality, we can assume $P_{01} \neq Q_{01}$, and choose α such that $0 < \alpha < \frac{P_{01} - Q_{01}}{2}$. The nodes i

and j are predicted to be in the same community if $\widehat{P}_{01}(i,j) > \frac{P_{01}+Q_{01}}{2}$, and the probability of making an error is

$$\mathbb{P}\left(\left|\widehat{P}_{01}(i,j) - P_{01}(i,j)\right| \geq \alpha\right).$$

By the union bound, the probability that all nodes are correctly classified is bounded by

$$\frac{n(n-1)}{2} \max_{ij} \mathbb{P}\left(\left|\widehat{P}_{01}(i,j) - P_{01}(i,j)\right| \geq \alpha\right),$$

where the maximum is taken over all nodes pair ij. By equation (6.19), for all node pairs ij, we have $\mathbb{P}\Big(|\widehat{P}_{01}(i,j) - P_{01}(i,j)| \geq \alpha\Big) \to 0$. Therefore, all nodes are a.s. correctly classified as $T \to \infty$. □

Remark 6.3. If P and Q are unknown, we can add a step, where the estimated transition matrices, $\widehat{P}(i,j)$, are clustered into two classes (for example using k-means).

6.3 Markovian Evolution of Community Memberships

This section focuses on clustering temporal networks, whose membership structure follows a Markov chain, but the interaction structure is time independent. Specifically, we denote by $z_{it} \in [K]$ the group membership of node i at time t. Then, across nodes, the random variables $(z_{it})_{1 \leq t \leq T}$ are independent and identically distributed. For each node i, the group membership $z_{i\cdot} = (z_{i1}, \cdots, z_{iT})$ follows an irreducible and aperiodic Markov chain, given by

$$\mathbb{P}(z_{i\cdot}) = \alpha_{z_{i1}} \prod_{t=2}^{T} \pi_{z_{i,t-1}, z_{it}}, \tag{6.20}$$

where α is the initial distribution and π is the transition probability matrix. Conditioned on the node labels, the edges are independent, and for all $i < j$ and all t, we have

$$A_{ij}^t \mid z_{it}, z_{jt} \sim \text{Ber}\left(p_{z_{it} z_{jt}}\right).$$

The likelihood of the sequence of adjacency matrices $A^{1:T} = \left(A^1, \cdots, A^T\right)$ is therefore

$$\mathbb{P}\left(A^{1:T} \mid Z\right) = \prod_{i=1}^{n} \alpha\left(z_{i1}\right) \prod_{t=2}^{T} \pi_{z_{i,t-1}, z_{it}} \prod_{t=1}^{T} \prod_{i<j} p_{z_{it} z_{jt}}^{A_{ij}^t} \left(1 - p_{z_{it} z_{jt}}\right)^{1 - A_{ij}^t}.$$

6.3.1 Variational Expectation-Maximization Algorithm

Let us first assume that K is known, and that α is the stationary distribution of π. We aim at estimating the group memberships $Z = (z_{it})_{1 \leq i \leq n, 1 \leq t \leq T}$ as well as the model parameters $\theta = (\pi, P)$ where $P = \left(p_{k\ell}\right)_{k,\ell \in [K]}$.

While the global maximisation of the likelihood is intractable when n or T are large, the Expectation-Maximization (EM) algorithm (Dempster *et al.*, 1977) provides a way to find local maxima. EM algorithm computes the conditional distribution of Z given the observation $A^{1:T}$. However, in our case this distribution does not factor into a product over the n nodes because of dependencies. Indeed, we have

$$\mathbb{P}\left(Z \mid A^{1:T}\right) = \mathbb{P}\left(z_{\cdot 1} \mid A^1\right) \prod_{t=2}^{T} \mathbb{P}\left(z_{\cdot t} \mid z_{\cdot t-1}, A^t\right),$$

where $z_{\cdot t} = (z_{1t}, \cdots, z_{nt})$ denotes the community labels at time t. Unfortunately, the distribution $\mathbb{P}\left(Z^t \mid Z^{t-1}, A^t\right)$ cannot be further factored as the random variables $z_{it} \mid A_{ij}^t$ and $z_{jt} \mid A_{ij}^t$ are not independent. Indeed, by observing an edge between i and j at time t the likelihood that $z_{it} = z_{jt}$ increases. The *variational approximation* introduces a class of probability distributions $\mathbb{Q}$ such that

$$\mathbb{Q}_\tau(Z) = \prod_{i=1}^{n} \mathbb{Q}_\tau\left(z_{i\cdot}\right) = \prod_{i=1}^{n} \mathbb{Q}_\tau\left(z_{i1}\right) \prod_{t=2}^{T} \mathbb{Q}_\tau\left(z_{it} \mid z_{it-1}\right).$$

We introduce $\tau(i,k) = \mathbb{Q}\left(z_{i1} = k\right)$ and $\tau(t,i,k,\ell) = \mathbb{Q}\left(z_{it} = \ell \mid z_{it-1} = k\right)$. Thus, under $\mathbb{Q}$, the distribution of $(z_{i1}, \ldots, z_{iT})$ is a time-inhomogeneous Markov chain with transitions $\tau(t,i,k,\ell)$ and initial distribution $\tau(i,k)$. In particular, $\sum_{k=1}^{K} \tau(i,k) = 1$ and $\sum_{\ell=1}^{K} \tau(t,i,k,\ell) = 1$ and

$$\mathbb{Q}(Z) = \prod_{i=1}^{n} \prod_{k=1}^{K} \tau(i,k)^{1(z_{it}=k)} \prod_{t=2}^{T} \prod_{1 \leq k,\ell,K} \tau(t,i,k,\ell)^{1(z_{it-1}=k)1(z_{it}=\ell)}.$$

The marginal distribution $\tau_{\text{marg}}(t, i, k) = \mathbb{Q}\left(z_{it} = k\right)$ are computed recursively by

$$\begin{aligned}\tau_{\text{marg}}(t, i, k) &= \tau(i, k), \\ \tau_{\text{marg}}(t, i, k) &= \sum_{\ell=1}^{K} \tau_{\text{marg}}(t-1, i, \ell)\tau(t, i, \ell, k).\end{aligned}$$

Variational Expectation-Maximization (VEM) algorithm (Matias and Miele, 2017) then seeks to maximise

$$J(\theta, \tau) = \mathbb{E}_{\mathbb{Q}}\left(\log \mathbb{P}\left(A^{1:T}, Z\right)\right) + \mathcal{H}(\mathbb{Q}),$$

where $\mathcal{H}(\mathbb{Q})$ denotes the entropy of $\mathbb{Q}$. Hence, $J(\theta, \tau)$ is equal to

$$\begin{aligned}&\sum_{i=1}^{n}\sum_{k=1}^{K} \tau(i, k)\left[\log \alpha_k - \log \tau(i, k)\right] \\ &\quad + \sum_{t=2}^{T}\sum_{i=1}^{n}\sum_{1\le k,\ell\le K} \tau_{\text{marg}}(t-1, i, k)\tau(t, i, k, \ell) \\ &\quad \times \left[\log \pi_{k\ell} - \log \tau(t, i, k, \ell)\right] \\ &\quad + \sum_{t=1}^{T}\sum_{1\le i<j\le n}\sum_{1\le k,\ell\le K} \tau_{\text{marg}}(t, i, k)\tau_{\text{marg}}(t, j, \ell) \\ &\quad \times \log\left(\text{Ber}\left(p_{z_{it}z_{jt}}\right)\left(A_{ij}^t\right)\right),\end{aligned}$$

with

$$\text{Ber}\left(p_{z_{it}z_{jt}}\right)\left(A_{ij}^t\right) = \begin{cases} p_{z_{it}z_{jt}} & \text{if } A_{ij}^t = 1, \\ 1 - p_{z_{it}z_{jt}} & \text{otherwise.}\end{cases}$$

The optimisation is done iteratively. At step k, with current estimates $\left(\tau^k, \theta^k\right)$, we perform the two following sub-steps:

1. **VE-step**: compute $\tau^{k+1} = \arg\max_\tau J\left(\theta^k, \tau\right)$;
2. **M-step**: compute $\theta^{k+1} = \arg\max_\theta J\left(\theta, \tau^{k+1}\right)$.

The following lemma provides the value of the updates τ^{k+1} and θ^{k+1}.

Lemma 6.3. *The value* $\widehat{\tau} = \arg\max_{\tau} J(\theta, \tau)$ *verifies*

$$\widehat{\tau}(t, i, k, \ell) \propto \pi_{k\ell} \prod_{j=1}^{n}\prod_{k'=1}^{K}\left(\text{Ber}\left(p_{z_{it}z_{jt}}\right)\left(A_{ij}^t\right)\right)^{\widehat{\tau}_{\text{marg}}(t,j,k')},$$

where the proportionality insures the normalisation constraints on τ. Similarly, $\widehat{\theta} = \arg\max_{\theta} J(\tau, \theta)$ is given by $\widehat{\theta} = \left(\widehat{\pi}, \widehat{P}\right)$ such that

$$\widehat{\pi}_{k\ell} \propto \sum_{t=2}^{T}\sum_{i=1}^{n} \tau_{\text{marg}}(t-1,i,k)\tau(t,i,k,\ell),$$

$$\widehat{p}_{k\ell} = \frac{\sum_{t=1}^{T}\sum_{1\le i,j\le n} \tau_{\text{marg}}(t,i,k)\tau_{\text{marg}}(t,i,\ell)1\left(A_{ij}^{t} \neq 0\right)}{\sum_{t=1}^{T}\sum_{1\le i,j\le n} \tau_{\text{marg}}(t,i,k)\tau_{\text{marg}}(t,i,\ell)}.$$

Proof. The proof follows from a direct derivation of $J(\tau, \theta)$. For example, we have

$$\begin{aligned}\frac{\partial J}{\partial \tau(t,i,k,\ell)} &= \tau_{\text{marg}}(t-1,i,k)\left[\log \pi_{k\ell} - \log \tau(t,i,k,\ell) + 1\right] \\ &\quad + \tau(t,i,k,\ell)\tau_{\text{marg}}(t-1,i,k)\log\left(\text{Ber}\left(p_{z_{it}z_{jt}}\right)\left(A_{ij}^{t}\right)\right)\end{aligned}$$

and equating this derivative to zero leads to the stated expression for $\widehat{\tau}(t,i,k,\ell)$. □

Finally, α is obtained by computing the empirical mean of the distribution $\widehat{\tau}_{\text{marg}}$ over all data points, *i.e.*,

$$\forall k \in [K] : \alpha_k = \frac{1}{nT}\sum_{t=1}^{T}\sum_{i=1}^{n} \widehat{\tau}_{\text{marg}}(t,i,k).$$

6.3.2 Belief Propagation Using the Space-time Graph

While the maximum likelihood estimator finds the membership structure that maximises the likelihood by solving $\arg\max_Z \mathbb{P}\left(A^{1:T} \mid Z\right)$, we will here instead try to find, for every node, the block assignment that maximises its *marginal likelihood*. More precisely, the marginal likelihood $\psi_k^i(t)$ is the probability that node i belongs at time t to block k according to the posterior distribution, and is given by

$$\psi_k^i(t) = \mathbb{P}\left(z_{it} = k \mid A^{1:T}\right).$$

Then, for every time t we will assign node i to be in block $\hat{z}_{it}$ such that

$$\hat{z}_{it} = \arg\max_{k\in[K]} \psi_k^i(t).$$

To compute the marginals, we model as if every neighbors j of a given node i at time t sends a message $\psi_k^{i\to j}(t)$, which is an estimate of the probability that i is in community k if node j was not here.

Since the graph is temporal, we also have to take into account the temporal evolution. More precisely, at time t each node i receives a message from its past and future copies, denoted by $\psi^{i(t-1)\to i(t)}$ and $\psi^{i(t)\to i(t+1)}$.

The update equation for the spatial messages is

$$\psi_k^{i\to j}(t) \propto \left(\sum_\ell \pi_{k\ell}\psi_\ell^{i(t-1)\to i(t)}\right)\left(\sum_\ell \pi_{\ell k}\psi_\ell^{i(t+1)\to i(t)}\right) \times \prod_{\substack{j:\,A_{ij}^t=1\\ j\neq i}} \sum_\ell p_{k\ell}\psi_\ell^{j\to i}(t)$$

where the proportionality hides a factor imposing the normalisation condition $\sum_k \psi_k^{i\to j} = 1$. The update equation omits the non-edges, as in sparse networks they can be approximated as a global interaction. Moreover, the update equation for $\psi_k^{i\to j}(t)$ does not involve the message that j sends to i, to avoid any "echo chamber" effect, where information would be amplified between i and j in a noisy fashion (for more details see Moore, 2017). In the similar fashion, the update equation for temporal messages is given by

$$\psi_k^{i(t)\to i(t+1)} \propto \left(\sum_\ell \pi_{k\ell}\psi_\ell^{i(t-1)\to i(t)}\right) \prod_{j:\,A_{ij}^t=1} \sum_\ell p_{k\ell}\psi_\ell^{j\to i}(t)$$

and a similar expression also holds for $\psi_k^{i(t-1)\to i(t)}$.

Belief propagation consists of initializing the messages randomly and then repeatedly updating them with the update equations. This is typically done asynchronously, by first choosing a node i and a time t uniformly at random and updating $\psi_k^{i\to j}(t)$ for all j and k, as well as $\psi_k^{i(t)\to i(t+1)}$ and $\psi_k^{i(t-1)\to i(t)}$. When convergence occurs, we compute the marginal of each vertex using

$$\psi_k^i(t) \propto \left(\sum_\ell \pi_{k\ell}\psi_\ell^{i(t-1)\to i(t)}\right)\left(\sum_\ell \pi_{k\ell}\psi_\ell^{i(t+1)\to i(t)}\right) \times \prod_{j:\,A_{ij}^t=1} \sum_\ell p_{k\ell}\psi_\ell^{j\to i}(t).$$

We finish this section by noticing that when $\pi = rI_K + \frac{1-r}{K}1_K^T 1_K$, we have

$$\sum_\ell \pi_{k\ell}\psi_\ell^{i(t-1)\to i(t)} = r\psi_k^{i(t-1)\to i(t)} + \frac{1-r}{K},$$

$$\sum_\ell \pi_{k\ell}\psi_\ell^{i(t+1)\to i(t)} = r\psi_k^{i(t+1)\to i(t)} + \frac{1-r}{K},$$

which further simplifies the update equations. Furthermore, in a homogeneous model ($p_{k\ell} = p_{\rm in}$ if $k = \ell$ and $p_{\rm out}$ otherwise), we have

$$\sum_{\ell} p_{k\ell} \psi_{\ell}^{j\to i}(t) = \lambda \psi_{k}^{j\to i}(t) + \frac{1-\lambda}{K}.$$

6.3.3 Online Inference as a Semi-supervised Problem

The lagging problem

In a framework where community memberhips vary with time, clustering by applying directly the time-aggregated spectral methods derived in Section 6.2.3 would fail. Indeed, time-varying community memberships lead to a contamination of the information given by the past interactions. For example, if node i changes its community assignment at time t_1, then one should not use the interactions of node i during the first t_1 snapshots to find its community membership at time $t > t_1$. This *lagging problem* especially complicates the situation when the layers are temporally correlated. To avoid this issue, we propose an online recovery of the node labels. More specifically:

- at time $t = 1$, we use a static community detection algorithm to output $\hat{z}_{\cdot 1} = (\hat{z}_{11}, \cdots, \hat{z}_{n1})$, a prediction of the initial node labels $z_{\cdot 1} = (z_{11}, \cdots, z_{n1})$ from the observation of the first snapshot A^1;
- at time $t > 1$, we will use the observation of the first t snapshots $A^1, \dots, A^t$ as well as the previous predictions $\hat{z}_{\cdot 1}, \cdots \hat{z}_{\cdot t-1}$. This will be treated as a semi-supervised learning problem, where the prediction $\hat{z}_{\cdot t-1}$ done at the previous time step is seen as a noisy oracle for the true node labelling $z_{\cdot t}$ at time t.

From the Markov structure, the prediction at time $t > 1$ reduces to predicting $z_{\cdot t}$ using only the network at time $t-1$ and t and the previous prediction $\hat{z}_{\cdot t-1}$. This can be interpreted as a noisy semi-supervised problem with oracle (see Section 5.4), where the previous prediction $\hat{z}_{\cdot t-1}$ plays the role of the oracle information for the node labels at time t. This oracle is noisy, as it bears two kinds of potential mistakes. Firstly, $\hat{z}_{\cdot t-1}$ is not necessarily exactly equal to the perfect community labelling $z_{\cdot t-1}$. Secondly, since the node labels vary through time, $z_{\cdot t-1}$ does not precisely correspond to $z_{\cdot t}$.

6.3.4 Degree-corrected Temporal SBM with Markov Community Memberships

In addition to the Markov community structure described in (6.20), we will assume for simplicity that the initial labels and the transitions are uniform, that is

$$\alpha = \frac{1}{K} 1_K \quad \text{and} \quad \pi = \eta I_K + \frac{1-\eta}{K} 1_K 1_K^T.$$

In other words, a node keeps its label with probability $\eta \in [0, 1]$, and choose a label uniformly at random with probability $1 - \eta$.

We then assume that the pair interaction between two nodes i and j is a Markov process depending only on the community labelling and on some degree correction parameters $\theta = (\theta_1, \cdots, \theta_N)$. In particular,

$$\mathbb{P}(A \,|\, z, \theta) = \prod_{1 \le i < j \le N} \mathbb{P}\left(A_{ij}^1 \,|\, z_{i1}, z_{j1}, \theta_i, \theta_j\right) \prod_{t=2}^{T} \mathbb{P}\left(A_{ij}^t \,|\, A_{ij}^{t-1}, z_{it}, z_{jt}, \theta_i, \theta_j\right).$$

We further consider a homogeneous model in which the initial distribution is given by

$$\mathbb{P}\left(A_{ij}^1 \,|\, z_{i1}, z_{j1}, \theta_i, \theta_j\right) = \begin{cases} \mu^{\theta_i\theta_j}\left(A_{ij}^1\right), & \text{if } z_{i1} = z_{j1}, \\ \nu^{\theta_i\theta_j}\left(A_{ij}^1\right), & \text{otherwise,} \end{cases}$$

and the transition probabilities are given by

$$\mathbb{P}\left(A_{ij}^t = b \,|\, A_{ij}^{t-1} = a, z_{it}, z_{jt}, \theta_i, \theta_j\right) = \begin{cases} P_{ab}^{\theta_i\theta_j}, & \text{if } z_{it} = z_{jt}, \\ Q_{ab}^{\theta_i\theta_j}, & \text{otherwise.} \end{cases}$$

Similarly to Section 6.2.3, the degree-corrected initial distributions are defined by

$$\mu^{\theta_i\theta_j} = \begin{pmatrix} 1 - \theta_i\theta_j\mu_1 \\ \theta_i\theta_j\mu_1 \end{pmatrix}, \quad \nu^{\theta_i\theta_j} = \begin{pmatrix} 1 - \theta_i\theta_j\nu_1 \\ \theta_i\theta_j\nu_1 \end{pmatrix},$$

and the transition probability matrices are given by

$$P^{\theta_i\theta_j} = \begin{pmatrix} 1 - \theta_i\theta_j P_{01} & \theta_i\theta_j P_{01} \\ 1 - P_{11} & P_{11} \end{pmatrix}, \quad Q^{\theta_i\theta_j} = \begin{pmatrix} 1 - \theta_i\theta_j Q_{01} & \theta_i\theta_j Q_{01} \\ 1 - Q_{11} & Q_{11} \end{pmatrix},$$

with the assumption $\min_{i,j}\{\theta_i\theta_j\delta\} \le 1$, where $\delta = \max\{\mu_1, \nu_1, P_{01}, Q_{01}\}$. We normalise the degree correction parameters so that for all k it holds that $\sum_i 1(z_{i1} = k)\theta_i = \sum_i 1(z_{i1} = k)$. Finally, we suppose that the transition probabilities and the degree-correction parameters do not vary with time, to avoid any parameter identifiability issues (Matias and Miele, 2017).

Online Maximum A Posteriori estimator

The following proposition gives the expression of the MAP estimator for the presented online learning problem.

Proposition 6.3. *Let $s \in [K]^n$ be a noisy oracle on the node labels at time t, which is supposed to be independent of the observed interactions A. Define the rate of mistake of s as $\rho = \mathbb{P}(s_i \neq \hat{z}_{it})$ and assume this rate is the same for all nodes. A Maximum A Posteriori estimator for the online learning problem described above is defined by*

$$\hat{z}_{\cdot t} = \arg\max_{z \in [K]^n} \mathbb{P}\left(z \mid A^t, A^{t-1}, s\right)$$

and is any labelling $z \in [K]^n$ that maximises

$$\sum_{\substack{i,j \\ z_i = z_j}} \left\{ \ell_{01}^{\theta_i\theta_j} \left(A_{ij}^t - A_{ij}^{t-1} A_{ij}^t\right) + \ell_{10}^{\theta_i\theta_j} \left(A_{ij}^{t-1} - A_{ij}^{t-1} A_{ij}^t\right) + \ell_{11}^{\theta_i\theta_j} A_{ij}^{t-1} A_{ij}^t \right.$$

$$\left. - \log \frac{Q_{00}^{\theta_i\theta_j}}{P_{00}^{\theta_i\theta_j}} \right\} + 2\lambda \sum_{i=1}^n 1\left(z_i = s_i\right),$$

where $\ell_{ab}^{\theta_i\theta_j} = \log \frac{P_{ab}^{\theta_i\theta_j}}{P_{ab}^{\theta_i\theta_j}} - \log \frac{P_{00}^{\theta_i\theta_j}}{P_{00}^{\theta_i\theta_j}}$ *and* $\lambda = \log \frac{1-\rho}{\rho}$.

Proof. By Bayes' rule,

$$\mathbb{P}\left(z \mid A^t, A^{t-1}, s, \theta\right) \propto \mathbb{P}\left(A^t \mid A^{t-1}, z, s, \theta\right) \mathbb{P}\left(z \mid A^{t-1}, s, \theta\right),$$

where the proportionality symbol hides a term $\mathbb{P}\left(A^t \mid A^{t-1}, s, \theta\right)$ independent of z. Since $\mathbb{P}\left(A^t \mid A^{t-1}, z, s, \theta\right) = \mathbb{P}\left(A^t \mid A^{t-1}, z, \theta\right)$, then proceeding similarly to the proof of Proposition 6.2, the log-likelihood term $\log \mathbb{P}\left(A^t \mid A^{t-1}, z, \theta\right)$ can be rewritten as

$$\frac{1}{2} \sum_{\substack{i,j \\ z_i = z_j}} \left\{ \ell_{01}^{\theta_i\theta_j} \left(A_{ij}^t - A_{ij}^{t-1} A_{ij}^t\right) + \ell_{10}^{\theta_i\theta_j} \left(A_{ij}^{t-1} - A_{ij}^{t-1} A_{ij}^t\right) \right.$$

$$\left. + \ell_{11}^{\theta_i\theta_j} A_{ij}^{t-1} A_{ij}^t - \log \frac{Q_{00}^{\theta_i\theta_j}}{P_{00}^{\theta_i\theta_j}} \right\}.$$

The oracle information is equal to

$$\begin{aligned} \mathbb{P}(z \mid s) &= \prod_{i=1}^n \frac{\mathbb{P}(s_i \mid z_i)}{\mathbb{P}(s_i)} \mathbb{P}(z_i) \\ &= (1-\rho)^{|\{i \in [n] \,:\, z_i = s_i\}|} \rho^{|\{i \in [n] \,:\, z_i \neq s_i\}|} \left(\frac{1}{K}\right)^n \\ &= \left(\frac{\rho}{1-\rho}\right)^{|\{i \in [n] \,:\, z_i \neq s_i\}|} (1-\rho)^n \left(\frac{1}{K}\right)^n \end{aligned}$$

where we used the uniformity of the node labels. □

Continuous relaxation of the MAP

For simplicity of the derivations to come, in this section we restrict the study to $K = 2$.

Denote by $A^t_{\text{pers}} = A^{t-1} \odot A^t$ the adjacency matrix corresponding to *persistent edges*, by $A_{\text{new}} = A^t - A^t_{\text{pers}}$ the adjacency matrix corresponding to *freshly formed* edges, and by $A_{\text{old}} = A^{t-1} - A_{\text{pers}}$ the adjacency matrix corresponding to *disappearing edges* between time $t-1$ and t. Then, using the Taylor expansion as in Section 6.2.3, we can approximate the MAP estimator by

$$\underset{z \in \{-1,1\}^n}{\arg\min} \; -z^T \left(W - \tau \frac{dd^T}{2m} \right) z + \lambda (s-z)^T (s-z) \tag{6.21}$$

where $W^t = \alpha_{01} A^t_{\text{new}} + \alpha_{10} A^t_{\text{old}} + \alpha_{11} A^t_{\text{pers}}$ with $\alpha_{ab} = \log \frac{P_{ab}}{Q_{ab}}$, τ is a resolution parameter, $d_i = \sum_{j=1}^n W^t_{ij}$, and $m = \frac{1}{2} \sum_{i=1}^n d_i$.

This minimisation problem is analogous to the one studied in Section 5.4 for noisy semi-supervised clustering in the DC-SBM. We can also propose the following continuous relaxation

$$\hat{x} = \underset{\substack{x \in \mathbb{R}^n \\ x^T D x = 2m}}{\arg\min} \; -x^T M x + \lambda (s-x)^T (s-x),$$

where $D = \operatorname{diag}(d_1, \cdots, d_n)$ and $M = W - \tau \frac{dd^T}{2m}$. The solution of this relaxation is determined by mimicking the reasoning of Section 5.4.2. In particular, by denoting the eigendecomposition of $D^{-1/2} (-M + \lambda I_n) D^{-1/2}$ by

$$D^{-1/2} (-M + \lambda I_n) D^{-1/2} = Q \Delta Q^T$$

with $\Delta = \operatorname{diag}(\delta_1, \ldots, \delta_n)$ and $QQ^T = I_n$ and letting $b = \lambda Q^t s$, we obtain that $\hat{x}$ verifies

$$(-M + \lambda I_n - \gamma_* D) \hat{x} = \lambda s, \tag{6.22}$$

where γ_* is the smallest solution of the *explicit secular equation* (Gander *et al.*, 1989)

$$\sum_{i=1}^n \left(\frac{b_i}{\delta_i - \gamma} \right)^2 - 2m = 0. \tag{6.23}$$

This leads to Algorithm 19.

Algorithm 19: Online clustering of time-varying communities.

Input: Observed graph sequence $A^{1:T} = (A^1, \dots, A^T)$; number of communities K; static graph clustering algorithm algo; parameters $\alpha_{01}, \alpha_{10}, \alpha_{11}$ and $\lambda_1, \dots, \lambda_T$.

Output: Node labelling $Z = (z_{it})$.

Initialize: Compute $\hat{z}_{\cdot t} \leftarrow \mathsf{algo}\left(A^1\right)$.

for $t = 2, \dots, T$ **do**

  Compute $W = \alpha_{01} A^t_{\text{new}} + \alpha_{10} A^t_{\text{old}} + \alpha_{11} A^t_{\text{pers}}$.

  Compute $M = W - \frac{dd^T}{2m}$ where $d_i = \sum_{j=1}^n W_{ij}$ and $m = \frac{1}{2}\sum_{i=1}^n d_i$.

  Let γ^* be the smallest solution of Equation (6.23).

  Compute $\hat{x}$ as the solution of Equation (6.22).

  Let $\hat{z}_{\cdot t} = \text{sign}(\hat{x})$.

Numerical experiments

We compare in Figure 6.6 the averaged accuracy obtained by Algorithm 19 with Algorithm 17 (spectral clustering with persistent edges) and an algorithm performing spectral clustering on each snapshot individually. In particular, we observe that when $\eta = 1$ (*i.e.*, static community structure), Algorithm 17 is extremely efficient, as expected. Since it takes into account all previous snapshots, it in particular outperforms Algorithm 19. On the contrary, when $\eta \neq 1$, the lagging problem arises, and Algorithm 17 ends up with a very poor accuracy after a few snapshots. On the contrary, Algorithm 19 keeps a very high accuracy over all snapshots.

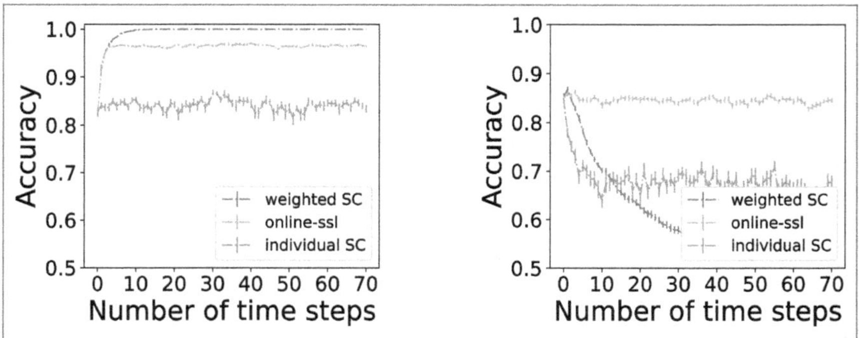

(a) $\eta = 1$ **(b)** $\eta = 0.85$

Figure 6.6. Accuracy of Algorithm 19 (*online-ssl*) with $\alpha_{01} = 1, \alpha_{10} = 0$ and $\alpha_{11} = 2$, on time-varying Markov Block Models with 300 nodes and $K = 2$ blocks (with uniform prior), and a stationary Markov edge evolution $\mu_1 = 0.05$, $\nu_1 = 0.02$, $P_{11} = 0.7$ and $Q_{11} = 0.3$. The results are averaged over 25 synthetic graphs, and error bars show the standard error. We compare with Algorithm 17 (*weighted SC* with $\alpha = 1, \beta = 2$) and an algorithm performing Spectral Clustering on each snapshot individually.

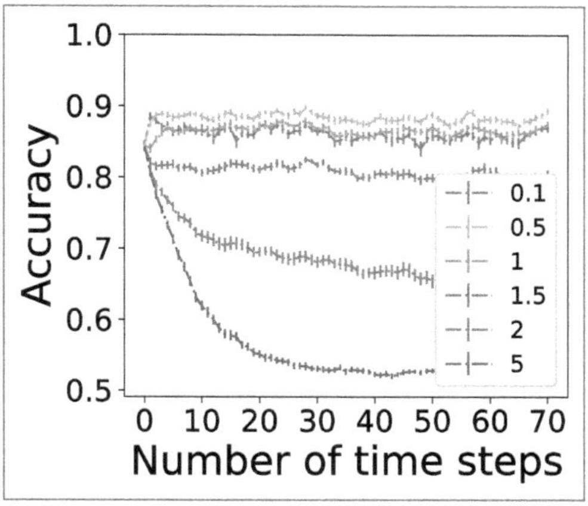

Figure 6.7. Accuracy of Algorithm 19 with $\alpha_{01} = 1$, $\alpha_{10} = 0$ and $\alpha_{22} = 2$ and various values of λ. Simulations are performed on time-varying Markov Block Models with $n = 300$, $K = 2$, $\mu_1 = 0.05$, $\mu_2 = 0.02$, $P_{11} = 0.7$, $Q_{11} = 0.3$ and $\eta = 0.9$. The results are averaged over 25 synthetic graphs, and error bars show the standard error.

In Figure 6.6, we choose λ_t to be constant and equal to 0.5, while Figure 6.7 explores other possible values. We observe that when λ_t is equal to a constant in the interval $[0.1, 1]$, Algorithm 19 outputs similar performances. On the other hand, when λ becomes too large, Algorithm 19 gives too much importance to the oracle, and the accuracy becomes worse. In practice, the choice of the parameters λ_t could be optimised from the data, *e.g.*, based on η or on the transition matrices P and Q. Moreover, it would be intuitive to increase λ_t with t, as the confidence in the oracle is higher when more temporal data is available. We leave this as a topic for future work.

Further Notes

We refer to Decelle *et al.*, 2011; Moore, 2017 for an extended description of the belief propagation techniques. Belief propagation was introduced for dynamic networks in Ghasemian *et al.*, 2016, and the extension for models incorporating link persistence is considered in Ghasemian, 2019. Similarly, Barucca *et al.*, 2018 studied a model with a Markov evolution of the community memberships and link persistence. While their interaction setting is restrictive, they showed that edge persistence increases the difficulty of community recovery.

Finally, some models also allow for an evolution of the interaction parameters over time (Xu and Hero, 2014; Bhattacharyya and Chatterjee, 2020). Nonetheless, it is important to note that identifiability issues often occur when both the memberships and the interaction kernels vary over time (Matias and Miele, 2017).

DOI: 10.1561/9781638280514.ch7

Chapter 7

Sampling in Networks

Many networks, including online and offline social networks, exist for which it is impossible to obtain a complete picture of the network. This leaves researchers with the need to develop sampling techniques for characterizing and studying large networks.

The general problem of sampling in a network can be formalised as follows. Let $G = (V, E)$ be an undirected network with $n = |V|$ nodes and $m = |E|$ links. We would like to design an efficient estimator of the average of a network function

$$\bar{f} = \frac{1}{n} \sum_{v \in V} f(v). \tag{7.1}$$

Despite the simplicity of the problem formulation, this can be used to describe many real-world statistical questions. Let us give just a few examples.

- How young is a social network? – Take as $f(v)$ the age of node v;
- How many friends on average has a social network member? – Take as $f(v)$ the degree (the number of friends) of node v;
- What is a proportion of a certain sub-population in a network? – Take $f(v) = 1$ if v belongs to that sub-population and otherwise $f(v) = 0$.

7.1 Overview of Sampling Methods

7.1.1 Independent Uniform Sampling

Clearly, the simplest unbiased estimator is the *independent uniform sampling* estimator. That is, obtain a set of samples $v_{i_1}, \ldots, v_{i_k}$, sampling each node independently with some probability p. Then,

$$\hat{f}^{(k)} = \frac{1}{k}\sum_{s=1}^{k} f(v_{i_s}). \tag{7.2}$$

Strictly speaking, here we consider sampling with replacement. However, in large networks hitting the same node twice occurs with a very small probability. This approach is widely attempted in practice but has at least the following two drawbacks: (i) in most cases, it is not easy at all to perform a uniform sampling. Take for example a questionary over the phone. If the phone numbers of stationary phones are mostly used, this can give an age-based bias. (ii) If one is interested in the study of a very small sub-population, it can be extremely difficult to collect enough samples from that sub-population. The latter concern has given motivation for the development of a number of methods based on the *chain-referral* approach, see e.g., Goodman, 1961.

7.1.2 Snowball Sampling

Snowball sampling is the first "naive" chain-referral approach. In a chain-referral approach, the sampling process starts from an initial subject (node), who provides one or several contacts of his/her friends (neighbours). Then, each new contact subject is approached for a questionary and then, after the questionary is completed, is asked to provide his/her contact list. This process continues until a sufficient number of samples is collected. The naive snowball estimator is very similar in its form to the estimator (7.2), namely

$$\hat{f}^{(k)} = \frac{1}{k}\sum_{s=1}^{k} f(v_{i_s}), \tag{7.3}$$

where $v_{i_1}, \ldots, v_{i_k}$ are contacted subjects, who gave answers.

Note that if at each stage we query just one neighbour from a contact list, this will correspond to a *random walk* on a social network.

One important problem with the naive snowball sampling is that the nodes with many neighbours are over-sampled (Erickson, 1979) because the random walk is more likely to come to such nodes.

7.1.3 Metropolis-Hastings Sampling

One natural way to mitigate the over-sampling of large degree nodes is to use the classical Markov chain techniques of Metropolis *et al.*, 1953–Hastings, 1970b.

To avoid the bias with respect to node degrees, we would like that the target distribution of the random walk will be uniform, *i.e.*, $\pi(v) = 1/n$. Then, according to the *Metropolis-Hastings* (MH) approach, we should change the probability of neighbour node selection to

$$\begin{aligned}\tilde{p}_{vu} &= \frac{1}{d(v)} \min\left\{1, \frac{\pi(u)p_{uv}}{\pi(v)p_{vu}}\right\} = \frac{1}{d(v)} \min\left\{1, \frac{d(v)}{d(u)}\right\} \\ &= \frac{1}{\max\{d(v), d(u)\}}\end{aligned}$$

if $(v, u) \in E$ and $v \neq u$, whereas $\tilde{p}_{vu} = 0$ if $(v, u) \notin E$ and $v \neq u$, and finally $\tilde{p}_{vu} = 1 - \sum_{s \neq v} \frac{1}{\max\{d(v),d(u)\}}$ if $u = v$. To summarise, we have

$$\tilde{p}_{vu} = \begin{cases} 0, & \text{if } (u, v) \notin E, \\ \frac{1}{\max\{d(v),d(u)\}}, & \text{if } (u, v) \in E \text{ and } v \neq u, \\ 1 - \sum_{s \neq v} \frac{1}{\max\{d(v),d(u)\}}, & \text{if } u = v. \end{cases} \tag{7.4}$$

Using the central limit theorem for Markov chains (see e.g., Brémaud, 1999), Avrachenkov *et al.*, 2018b established the asymptotic consistency for the estimator (7.3), where the samples $v_{i_1}, \ldots, v_{i_k}$ are generated according to (7.4). Specifically, we can state the following theorem.

Theorem 7.1 (Central Limit Theorem for MH-estimator). *For MH-estimator, it holds that*

$$\sqrt{k}\left(\hat{f}^{(k)} - \bar{f}\right) \xrightarrow{D} \mathcal{N}(0, \sigma^2_{MH}), \quad \text{as} \quad k \to \infty,$$

where $\sigma^2_{MH} = \frac{2}{n} f^T Z f - \frac{1}{n} f^T f - \left(\frac{1}{n} f^T \underline{1}\right)^2$, $f^T = (f(1), \ldots, f(n))$ *and where* $Z = [I - \tilde{P} + \frac{1}{n}\underline{1}\underline{1}^T]^{-1}$ *is the fundamental matrix.*

In the context of online social networks, the use of MH-estimator was first proposed by Gjoka *et al.*, 2010.

7.1.4 Respondent-driven Sampling

One important problem with MH-estimator is that it resamples many nodes and thus it is not very efficient. This problem can be corrected by the *Respondent-Driven Sampling* (RDS) proposed in a series of works by Heckathorn, 1997; Salganik and

Heckathorn, 2004; Volz and Heckathorn, 2008. In RDS the underlying sampling process is carried out with a standard random walk but the estimator is modified as follows:

$$\hat{f}^{(k)} = \frac{2m}{k}\sum_{s=1}^{k}\frac{f(v_{i_s})}{d(v_{i_s})}, \tag{7.5}$$

where $d(v_{i_s})$ is the degree (the number of neighbours) of node v_{i_s} and m is the number of links in the network. Of course, the value of m may be not available or difficult to estimate. This is mitigated in the following modification of the RDS-estimator

$$\hat{f}^{(k)} = \frac{\sum_{s=1}^{k} f(v_{i_s})/d(v_{i_s})}{\sum_{s=1}^{k} 1/d(v_{i_s})}. \tag{7.6}$$

The RDS-estimators (7.5) and (7.6) are asymptotically consistent and the corresponding CLTs can be found in Avrachenkov *et al.*, 2018b.

7.1.5 Respondent-driven Sampling with Uniform Jumps

The RDS-estimator still has the following problem: the random walk can be trapped in a sub-network with few connections to the other parts of the network. To overcome this problem, Avrachenkov *et al.*, 2010 suggested to combine the random walk with uniform jumps. Specifically, let us modify the network adjacency matrix A in the following way:

$$\tilde{A} = A + \frac{\alpha}{n}\underline{1}\underline{1}^T.$$

Namely, we add an artificial link with the weight α between any two nodes. One interpretation of this modification is that we combine the random walk based sampling with the uniform sampling.

Typically, the weight α is small, as one sample of the uniform sampling is more costly than one sample of the random walk based sampling. For example, in an online social network, where users are associated with unique numeric IDs, uniform node sampling is performed by querying randomly generated IDs. In practice, however, these samples are expensive (resource-wise) operations as the ID space in an OSN, such as Facebook and Myspace, is large and sparse. For instance, in Myspace only 10% of the IDs belong to valid users (Gauvin *et al.*, 2010), *i.e.*, only one in every ten queries successfully finds a valid Myspace account. In this example, a natural choice for the parameter α is 1/10.

Note that the random walk on the weighted graph defined by $\tilde{A}$ is still a random walk on an undirected graph and hence its stationary distribution is proportional to the weighted degree, *i.e.*,

$$\tilde{\pi}(v) = \frac{d(v)+\alpha}{2m+\alpha n} = \frac{1}{n}\frac{d(v)+\alpha}{\bar{d}+\alpha},$$

where $\bar{d}$ is the average degree of the network. Thus, we can modify the RDS-estimator as follows:

$$\hat{f}^{(k)} = \frac{1}{k}\sum_{s=1}^{k}\frac{f(v_{i_s})}{\tilde{\pi}(v_{i_s})} = \frac{n(\bar{d}+\alpha)}{k}\sum_{s=1}^{k}\frac{f(v_{i_s})}{d(v_{i_s})+\alpha}. \tag{7.7}$$

If the average degree and the total number of nodes are unknown, one can use a modification, similar to (7.6), *i.e.*,

$$\hat{f}^{(k)} = \frac{\sum_{s=1}^{k} f(v_{i_s})/(d(v_{i_s})+\alpha)}{\sum_{s=1}^{k} 1/(d(v_{i_s})+\alpha)}. \tag{7.8}$$

One more natural candidate for the combination of a random walk with uniform restart is the modification in PageRank style. That is, one can change the transition probability matrix of the random walk as follows:

$$\tilde{P} = (1-\epsilon)P + \epsilon\frac{1}{n}\underline{1}\underline{1}^T. \tag{7.9}$$

One big disadvantage of this approach is that even in the case of an undirected graph the stationary distribution of $\tilde{P}$, PageRank, does not have an explicit expression, which could be used in (7.7). Of course, one can then use Metropolis-Hastings modification of the transition probabilities. However, as we noted before, such modification leads to frequent resampling.

Interestingly, the random walk with uniform jumps defined by the modified adjacency matrix $\tilde{A}$ can be viewed as PageRank with node-dependent restart probability. To see this, we can transform the transition probability matrix of the random walk with jumps as follows:

$$\begin{aligned}\tilde{P} &= (D+\alpha I)^{-1}(A+\frac{\alpha}{n}\underline{1}\underline{1}^T)\\ &= (D+\alpha I)^{-1}DD^{-1}A + (D+\alpha I)^{-1}\alpha I\frac{1}{n}\underline{1}\underline{1}^T,\end{aligned}$$

which is the expression (3.8) with the restart probability matrix

$$C = (D+\alpha I)^{-1}D = \operatorname{diag}\left(\frac{d(i)}{d(i)+\alpha}\right)$$

and the personalization distribution $\nu = \frac{1}{n}\underline{1}^T$. Thus, $\tilde{\pi}(v)$ is the Occupation-Time Personalized PageRank (OT-PPR) of node i, defined in (3.9). In particular, the expression for C means that the random walk with jumps restarts with higher probability from large degree nodes.

Finally, the expressions (3.14) and (3.15) give us a useful formula for the expected time between consecutive restarts

$$\begin{aligned} E[\text{time between consecutive restarts}] &= \left(\sum_{i \in V}\left(1 - \frac{d(i)}{d(i)+\alpha}\right)\frac{d(i)+\alpha}{2m+\alpha n}\right)^{-1} \\ &= \frac{2m+\alpha n}{n\alpha} \\ &= \frac{\bar{d}+\alpha}{\alpha}. \end{aligned}$$

This formula allows us to tune the frequency of jumps by varying the parameter α.

7.1.6 Ratio with Tours Estimator

It may be difficult to do uniform sampling even from time to time. Therefore, instead of creating artificial links between all nodes, we can create artificial links between some nodes. Intuitively, it is beneficial to create artificial links between nodes from very different parts of the network. This should significantly increase the mixing time of the random walk. We can also consider the artificially linked nodes as one *super-node*. Let us denote the set of such nodes by S. Figure 7.1 illustrates the idea of the super-node. Note that now the graph can have multiple links and the transition probability for the random walk needs to be modified in proportion to the multiple links.

(a) Original network. **(b)** Modified network with a super-node, $S_4 = \{c, g, k, q\}$.

Figure 7.1. Construction of a super-node. This figure has appeared first in Avrachenkov *et al.*, 2016c.

Once the super-node is created, we can run the random walk in tours which start and end at the super-node, and use the following Ratio with Tours estimator (RT-estimator):

$$\hat{f}^{(k)} = \frac{\sum_{k=1}^{m(B)} \sum_{t=1}^{\xi_k-1} f(v_{i_t})/d(v_{i_t}) + 1/d_S \sum_{v\in S} f(v)}{\sum_{k=1}^{m(B)} \sum_{t=1}^{\xi_k-1} 1/d(v_{i_t}) + n/d_S}, \tag{7.10}$$

where ξ_k is the length of the k-th tour, B is the sampling budget, $m(B)$ is the number of tours until the budget is exhausted, i.e.

$$m(B) = \left\{k : \sum_{j=1}^{k} \xi_j \leq B\right\}, \tag{7.11}$$

d_S is the degree of the super-node, and

$$\tilde{f}(v) = \begin{cases} f(v), & \text{if } v \notin S, \\ 0, & \text{if } v \in S. \end{cases}$$

7.2 Tour-based Estimators for Motif Counting

Motif counting is an important task in network analysis. For instance, we need to count triangles and wedges to calculate the (global) clustering coefficient.

Cooper *et al.*, 2016 proposed tour based estimators for efficient estimation of network motifs. We note that their approach can be combined with the idea of super-node. To keep explanations transparent, let us consider tours of the random walk, which start and end at a single node, say s. Then, as before, let ξ_j denote the length of the j-th tour. Let π_s be the stationary probability of the random walk to be at node s. Then, we know that

$$E_s[\xi_j] = \frac{1}{\pi_s} = \frac{2m}{d_s}.$$

Thus, we can use the following estimator for the number of links:

$$\hat{m} = \frac{d_s}{2} \frac{1}{m(B)} \sum_{k=1}^{m(B)} \xi_k, \tag{7.12}$$

where $m(B)$ is defined in (7.11).

Next, if we want to estimate the number of triangles, we consider a random walk on a *weighted* network, where for each link $\{v, u\}$ we assign a weight $1 + t(\{v, u\})$, with $t(\{v, u\})$ being the number of triangles containing $\{v, u\}$.

The stationary distribution of the random walk on such weighted network is given by

$$\pi_v = \frac{d(v) + \sum_{u \in N(v)} t(\{v, u\})}{2m + 6t(G)},$$

where $t(G)$ is the number of triangles in the network.

Thus, we can use the following estimator for the number of triangles:

$$\hat{t} = \max\left\{0, \frac{(d(s) + \sum_{u \in N(s)} t(\{s, u\})) \sum_{k=1}^{m(B)} \xi_k}{6m(B)} - \frac{\hat{m}}{3}\right\},$$

where $\hat{m}$ is an estimate of the number of edges, e.g., given by (7.12).

It is straightforward to apply this approach to counting any network motif.

7.3 Numerical Comparison of Sampling Methods

7.3.1 Synthetic Networks

We first consider a SBM with $n = 20000$ nodes clustered in two communities of respective sizes 200 and 19800. We let $p_{11} = 0.3$, while $p_{12} = p_{22} = 0.001$. This models a small sub-population in a large social network. As the function to average, we first choose $f(v) = 1$, if node v is in the smallest cluster, and $f(v) = 0$, otherwise. The results are plot in Figure 7.2. We observe that uniform sampling provides excellent results, even using only $k = 500$ randomly chosen nodes, while the "naive" snowball sampling yields an over-estimation. This is expected since the standard random walk is biased towards large degree nodes, and in this situation

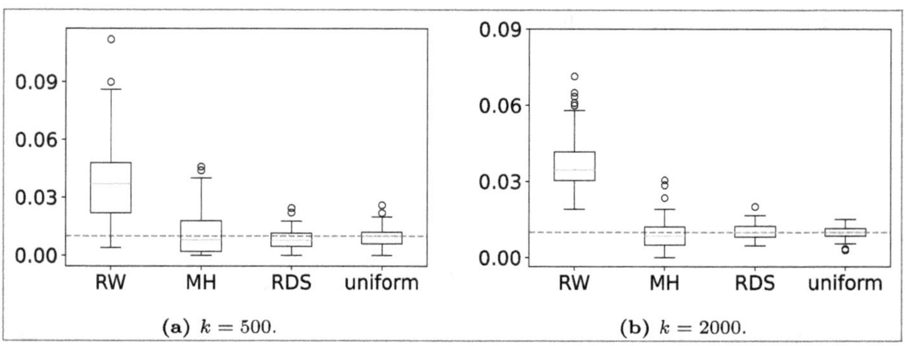

Figure 7.2. Different methods sampling the proportion of nodes in the smallest community of a SBM for a sampling budget of $k = 500$ and $k = 2000$. The two communities are of size 200 and 19800, and the probability of links are $p_{11} = 0.3$ while $p_{12} = p_{22} = 0.001$. The correct proportion is thus 0.01, and the boxplots show the results of 100 sampling trials.

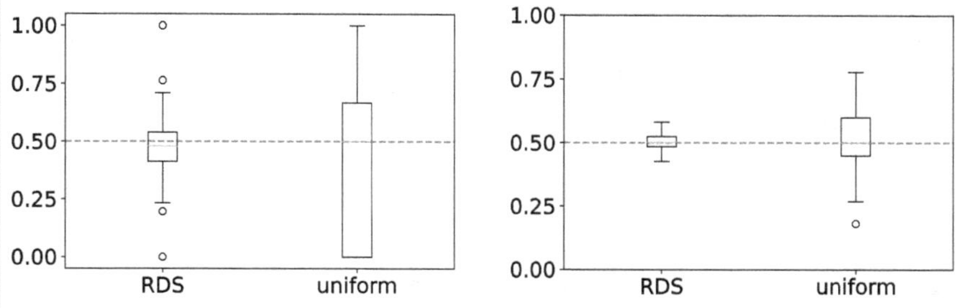

(a) $k = 300$. (b) $k = 2000$.

Figure 7.3. The performance of RDS and uniform sampling for estimating the proportion of nodes in Group-A in SBM for a sampling budget of $k = 300$ and $k = 2000$. The two communities are of size 500 and 49500, and the probability of links are $p_{11} = 0.8$ and $p_{12} = p_{22} = 0.0005$. Nodes in the largest community belong to Group-C, whereas the nodes in the smallest community are equally split into Group-A and Group-B. The correct proportion of node from Group-A with respect to nodes from both Group-A and Group-B is thus 0.5, and the boxplots show the results of 100 sampling experiments.

the large degree nodes are located in the smallest community. On the other hand, Metropolis-Hastings sampling and RDS successfully correct this bias.

To show why uniform sampling might not always perform best, we propose the following scenario. As before, we take a SBM with a large and a small community (sizes 49,500 and 500, respectively). We affect the nodes of the small community into two groups of equal sizes (called Group-A and Group-B). The nodes in the large community are all assigned to another Group-C. The goal is to recover the proportion of nodes in Group-A among the nodes in the small community. A practical motivation for this scenario could be that the small cluster represents a hard-to-reach sub-population, e.g., drug addicts. In this example the small community is further divided into the heavy users and the light users. One could be interested in the proportion of heavy users among the drug users. We assume that we know 10 nodes that belong to Group-A. We merge those nodes into a *super-node*, and perform RDS on this modified graph, which we compare with uniform sampling. The results are shown in Figure 7.3. We observe that RDS with super-node gives estimation with much less variance.

7.3.2 Real-world Network: DBLP

We will now compare different sampling methods on the *DBLP* data set ($n = 317,080$ nodes and $m = 1,049,866$ edges). In Figure 7.4, we estimate the average degree, *i.e.*, $f(v) = d(v)$. We also estimate the number of nodes with degree larger than 50 by considering $f(v) = 1(d(v) \geq 50)$ in Figure 7.5. In both cases, we observe that sampling using Metropolis-Hastings produces larger variance than RDS or uniform sampling.

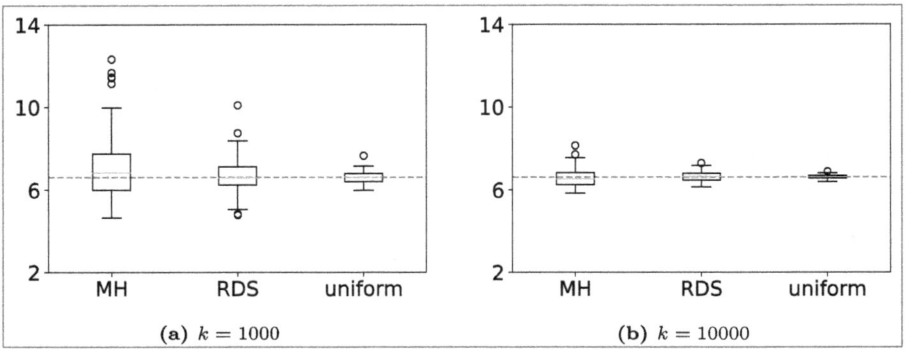

Figure 7.4. Estimation of the average degree of the *DBLP* data set using different methods, with budgets $k = 1000$ and $k = 10000$. The boxplot show the results of 100 sampling experiments. The correct value of the average degree is 6.6.

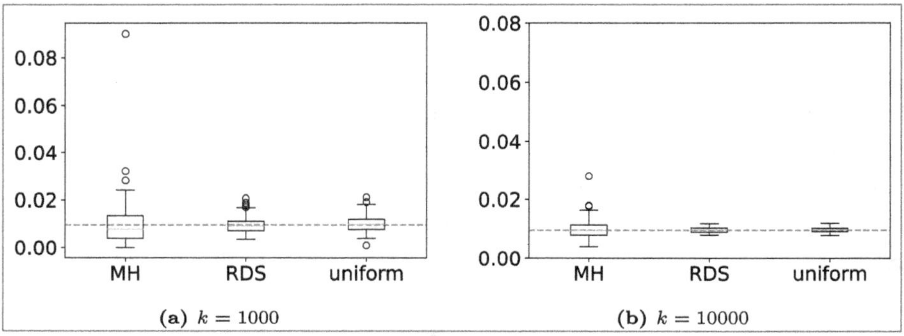

Figure 7.5. Estimation of the proportion of large degree nodes (defined as having a degree larger than 50) in the *DBLP* data set using different methods, with budgets $k = 1000$ and $k = 10000$. The boxplots show the results of 100 sampling experiments. The correct proportion is 0.01.

Further Notes

An interesting approach proposed by Dasgupta *et al.*, 2012, called *social sampling*, can be viewed as an intermediary between uniform node sampling and random walk based sampling. In this approach, once a node is sampled, the information about its neighbours also becomes available. Clearly, such an approach, if feasible, requires fewer samples than the uniform node sampling and avoids dependencies created by the random walk based methods.

It can be beneficial to sample a network using multiple random walks run in parallel, see Ribeiro and Towsley, 2010. To improve the efficiency, the multiple random walks should be either dependent in a special way or independent but timed as continuous random walks with transition rates proportional to the node degree.

As discussed by Avrachenkov *et al.*, 2016b, in certain cases, it can be beneficial to skip some samples in chain-referral methods. Intuitively, skipping some samples reduces correlation in random-walk based methods.

Instead of network functions defined over the nodes, one can consider network functions defined over the links or other motifs like triangles. For details on this, see Avrachenkov *et al.*, 2016c.

DOI: 10.1561/9781638280514.ch8

Appendix A

Background Material from Probability, Linear Algebra and Graph Theory

A.1 Probability

A.1.1 Probability Toolbox

In all the following, X (or X_i) denotes a random variable (r.v.).

Proposition A.1

- For a, b constants, $\mathbb{E}(aX + b) = a\mathbb{E}(X) + b$;
- $\mathbb{E}(X_1 + \cdots + X_m) = \mathbb{E}(X_1) + \cdots + \mathbb{E}(X_m)$;
- Let X be a r.v., A be an event and $1_A(X)$ be the indicator that event is realized by X. Then:

$$\mathbb{E}\big(1_A(X)\big) = \mathbb{P}(X \in A).$$

Definition A.1. The variance of a random variable X is given by

$$\mathrm{Var}(X) = \mathbb{E}\Big(\big(X - \mathbb{E}(X)\big)^2\Big).$$

Proposition A.2. *We have the following results:*

- $\mathrm{Var}(X) = \mathbb{E}(X^2) - (E(X))^2$;
- *for* a, b *constants,* $\mathrm{Var}(aX + b) = a^2 \mathrm{Var}(X)$;
- *if* $X_1, \ldots, X_m$ *are mutually independent, then* $\mathrm{Var}(X_1 + \cdots + X_m) = \mathrm{Var}(X_1) + \cdots + \mathrm{Var}(X_m)$;
- *if we do not have this independence, then* $\mathrm{Var}(X_1 + \cdots + X_m) = \mathrm{Var}(X_1) + \cdots + \mathrm{Var}(X_m) + \sum_{i \neq j} \mathrm{Cov}(X_i, X_j)$, *where* $\mathrm{Cov}(X_i, X_j) = \mathbb{E}(X_i X_j) - \mathbb{E}(X_i)\mathbb{E}(X_j)$.

A.1.2 Basic Probability Laws

Definition A.2. A random variable X is generated by a Bernoulli law with parameter $p \in [0; 1]$, denoted $X \sim \mathrm{Ber}(p)$, if:

1. X takes values in $\{0; 1\}$;
2. $\mathbb{P}(X = 1) = p$ and $\mathbb{P}(X = 0) = 1 - p$.

Example A.1. A r.v. $\mathrm{Ber}(p)$ models the result when we toss a biased coin (p is the probability of winning the coin toss).

Proposition A.3. *Let* $X \sim \mathrm{Ber}(p)$. *We have* $\mathbb{E}X = p$ *and* $\mathrm{Var}\, X = p(1 - p)$.

Definition A.3. The binomial distribution with parameters n and p, denoted $\mathrm{Bin}(n, p)$, is the discrete probability distribution of the number of successes in a sequence of n independent Bernoulli trials with parameter p.

Proposition A.4. *If* $(X_i)_{i=1,\ldots,n}$ *is a sequence of* n *i.i.d. random variable distributed according to* $\mathrm{Ber}(p)$, *then* $\sum_i X_i \sim \mathrm{Bin}(n, p)$.

Corollary A.1. *Let* $X \sim \mathrm{Bin}(n, p)$. *Then* $\mathbb{P}(X = k) = \binom{n}{k} p^k (1 - p)^{n-k}$. *Moreover,* $\mathbb{E}X = np$ *and* $\mathrm{Var}\, X = np(1 - p)$.

Definition A.4. The geometric distribution with parameter p, denoted $\mathrm{Geo}(p)$, is the probability of the number of Bernoulli trials (of parameter p) needed to get one success. In particular, if $X \sim \mathrm{Geo}(p)$, then $X \in \{1, 2, \cdots\}$ and $\mathbb{P}(X = k) = (1 - p)^{k-1} p$.

Proposition A.5. *Let* $X \sim \mathrm{Geo}(p)$. *Then* $\mathbb{E}X = \frac{1}{p}$ *and* $\mathrm{Var}\, p = \frac{1-p}{p^2}$.

A.1.3 Concentration of Random Variables

First moment inequalities

Proposition A.6 (Markov's inequality). *Let X be a random variable with positive values, and $a \in \mathbb{R}_+$. We have:*

$$\mathbb{P}(X \geq a) \leq \frac{\mathbb{E}X}{a}.$$

Proof. $\mathbb{E}X \geq \mathbb{E}\left(X 1_{X \geq a}\right) \geq a\mathbb{E}\left(1_{X \geq a}\right) = a\mathbb{P}\left(X \geq a\right).$ □

Remark A.1. By letting $a = t\mathbb{E}X$, we obtain $\mathbb{P}(X \geq t\mathbb{E}X) \leq \frac{1}{t}$. The convergence speed $\frac{1}{t}$ is rather slow, and depending on the requirements may not be strong enough.

Corollary A.2 (First moment method). *Let X be a positive, integer-valued random variable. We have:*

$$\mathbb{P}(X \neq 0) \leq \mathbb{E}(X).$$

The first moment is an upper bound on the probability that an integer random variable is not equal to zero.

Proof. Since X is integer valued, we have $\mathbb{P}(X \neq 0) = \mathbb{P}(X > 0) = \mathbb{P}(X \geq 1)$, and from there we can use Markov's inequality. □

Application A.3 (Union bound). Let $A_1, \cdots, A_m$ be a collection of events. Then,

$$\mathbb{P}(A_1 \cup \cdots \cup A_n) \leq \sum_{i=1}^{m} \mathbb{P}(A_i).$$

This can be shown by using the first moment method on $X = \sum_{i=1}^{m} 1_{A_i}$ and observing that $\{X > 0\} = A_1 \cup \cdots \cup A_m$.

Remark A.2. The first moment method is generally used when we have a sequence of integer, positive r.v. X_n such that $\mathbb{E}X_n \to 0$. In that case, $X_n \to 0$ almost surely.

We could naively imagine that, if $\mathbb{E}X_n \to +\infty$, then $\mathbb{P}(X_n > 0) \to 1$. Unfortunately, this is not true, and the next example provides a counter-example.

Example A.2. Let us take X_n such that $X_n = n^2$ with probability $1/n$ and $X_n = 0$ otherwise. Then, $\mathbb{E}(X_n) = n \to +\infty$, but $X_n \to 0$. Loosely speaking, this happens because the variance of X_n is very large. Indeed, $\operatorname{Var} X_n = n^2(n-1)$.

Second moment inequalities

Proposition A.7 (Chebyshev's inequality). *Let X be a random variable, and $a > 0$. We have:*

$$\mathbb{P}\Big(\big|X - \mathbb{E}X\big| \geq a\Big) \;\leq\; \frac{\operatorname{Var} X}{a^2}.$$

Proof. Apply Markov's inequality to $Y = (X - \mathbb{E}X)^2$. □

Example A.3. Let X be Gaussian $\mathcal{N}(0, \sigma^2)$. Then $\mathbb{E}|X| = \sigma\sqrt{\frac{2}{\pi}}$. Then Markov's inequality applied to $|X|$ gives

$$\mathbb{P}(X \geq a) \;\leq\; \sqrt{\frac{2}{\pi}}\frac{\sigma}{a},$$

while Chebyshev's inequality leads to

$$\mathbb{P}(X \geq a) \;\leq\; \left(\frac{\sigma}{a}\right)^2.$$

Chebyshev's inequality provides a stronger bound when a is large.

Application A.4 (Weak law of Large Numbers). Let $X_1, \ldots X_n$ be independent r.v. with mean μ and variance $\sigma^2 < +\infty$. Then:

$$\mathbb{P}\left(\left|\frac{X_1 + \cdots + X_n}{n} - \mu\right| > \epsilon\right) \to 0.$$

With some extra work, we can show that the condition $\sigma^2 < +\infty$ is not needed. Moreover, the strong law of large numbers states that the convergence holds in fact almost surely (and not simply in probability, as we have here).

Proof. Applying Chebychev's inequality for $U_n = \dfrac{X_1 + \cdots + X_n}{n}$, which has a mean μ and variance σ^2, leads to:

$$\mathbb{P}\big(|U_n| \geq \epsilon\big) \;\leq\; \frac{\sigma^2}{n\epsilon^2} \;\to\; 0.$$

□

Corollary A.5 (Second moment method). *Let X be a positive random variable. We have:*

$$\mathbb{P}\big(X = 0\big) \;\leq\; \frac{\operatorname{Var} X}{\big(\mathbb{E}X\big)^2} \;=\; \frac{\mathbb{E}(X^2)}{(\mathbb{E}X)^2} - 1.$$

Proof. We apply Chebychev's inequality with $a = \mathbb{E}X$:

$$\mathbb{P}(X = 0) \leq \mathbb{P}\Big(|X - \mathbb{E}X| \geq \mathbb{E}X\Big) \leq \frac{\operatorname{Var} X}{(\mathbb{E}X)^2},$$

where the first inequality holds since $|X - \mathbb{E}X| \geq \mathbb{E}X \Rightarrow X \leq 0$ or $X \geq 2\mathbb{E}X$. □

Remark A.3. From Cauchy-Schwarz inequality,

$$\mathbb{E}(X) \leq \mathbb{E}(X 1_{X>0}) \leq \sqrt{\mathbb{E}(X^2)}\sqrt{\mathbb{P}(X > 0)},$$

and thus $\mathbb{P}(X = 0) = 1 - \mathbb{P}(X > 0) \leq \dfrac{\operatorname{Var}(X)}{\mathbb{E}(X^2)}$, which provides a slightly stronger inequality than Corollary A.5.

Concentration of sums of i.i.d. random variables

Proposition A.8 (Hoeffding's inequality). *Let X_i be some independent random variables, such that $a_i \leq X_i \leq b_i$, and $S_n = \sum_{i=1}^n X_i$. For $t > 0$, we have:*

$$\mathbb{P}\Big(S_n \geq \mathbb{E}S_n + t\Big) \leq \exp\Big(-\frac{2t^2}{\sum_i (b_i - a_i)^2}\Big),$$

$$\mathbb{P}\Big(S_n \geq \mathbb{E}S_n - t\Big) \leq \exp\Big(-\frac{2t^2}{\sum_i (b_i - a_i)^2}\Big),$$

$$\mathbb{P}\Big(|S_n - \mathbb{E}S_n| \geq t\Big) \leq 2\exp\Big(-\frac{2t^2}{\sum_i (b_i - a_i)^2}\Big).$$

More details about concentration inequalities can be found for example in Vershynin, 2018, Chapter 2.

A.2 Graph Theory

A.2.1 Definitions, Vocabulary

Definition A.5. A *graph* G is a pair (V, E), where V is a finite set, whose elements are called *nodes* (or vertices, points) and E is a set of ordered node pairs called *edges* (or links, lines, bonds). Moreover, we use the following vocabulary:

- if $(ij) \in E \iff (ji) \in E$, then the graph is said to be *undirected* (this means that if there is a link going from i to j, there exists the same link in opposite direction);
- the edges (ii) are called *self-loops*. In particular, if for all nodes i, $(ii) \notin E$, we say that there is no self-loops;
- the graph is *weighted* if every edge $(ij) \in E$ has a weight $w_{ij} > 0$;

- we call *in-degree* of node i, denoted d_i^{in}, the number of (possibly weighted) edges coming to i, that is $d_i^{\text{in}} = \sum_{j \in V} w_{ji}$. Similarly, the *out-degree* of node i is the number of edges going from i, that is $d_i^{\text{out}} = \sum_{j \in V} w_{ij}$. For an undirected graph, $d_i^{\text{in}} = d_i^{\text{out}} = d_i$ and we simply call d_i the degree of node i.

Definition A.6.

- We call a path of G of length k a sequence $e_1, \ldots, e_k$ of edges $e_i = (v_{i-1}, v_i)$ where the v_i are vertices;
- A k-cycle is a path of legnth k that starts and ends at the same vertex;
- Suppose G is undirected. We say that two nodes u, v are connected if there exists a path going from u to v. We denote this as $u \leftrightarrow v$.

Proposition A.9. *The relation $\leftrightarrow$ is an equivalence relationship for the undirected graphs. In particular, we can partition the nodes into equivalent classes, called the* connected components.

Proof. We have $u \leftrightarrow u$ (path of length 0). Moreover, if $u \leftrightarrow v$ and $v \leftrightarrow z$, then $u \leftrightarrow z$ (by combining the two paths); this ensures transitivity. Finally, $u \leftrightarrow v$ implies $v \leftrightarrow u$ (the same path, on the opposite direction): this ensures symmetry. □

Remark A.4. In particular, this means that there exists a path between two nodes in a same connected component. Reciprocally, no path connects two nodes belonging to two different connected components.

Definition A.7. We say that G is *connected* if G has only one equivalent class under the relation $\leftrightarrow$. We say G is *disconnected* otherwise.

In particular, in a connected graph, for every node i and j, there exists a path going from i to j.

Definition A.8. Let i, j be two nodes. We call the distance between i and j, and denote $d(i,j)$ the length of the shortest path between i and j. If $i \not\leftrightarrow j$, then $d(i,j) = +\infty$.

Definition A.9 (Diameter)**.** We call diameter of a graph the largest distance between any pair of connected vertices.

A.2.2 Adjacency Matrix

Definition A.10. Let $G = (V, E)$ be an unweighted graph with n nodes. The adjacency matrix of G (denoted by A) is the binary matrix $A \in \{0, 1\}^{n \times n}$ such that $A_{ij} = 1$ if $(ij) \in E$.

We can easily extend this definition to a weighted graph: the element A_{ij} is then equal to the weight w_{ij} of the edge between nodes i and j.

Remark A.5. A is symmetric if and only if the graph is undirected. Moreover, the diagonal elements of A are zeros if and only if the graph does not have any self-loops.

Definition A.11. We call the degree matrix, denoted D, of a graph G the diagonal matrix whose diagonal element D_{ii} is the degree of node i.

A.2.3 Graph Laplacians

In the following, we will consider G as an undirected, weighted graph, on vertex set $V = \{1, \dots, n\}$. We denote by A the adjacency matrix of G, and by D its degree matrix.

Definition A.12 (Graph Laplacian)**.** We define:

- the *(standard or combinatorial) Laplacian* $L = D - A$;
- the *normalized Laplacian* $\mathcal{L} = D^{-1/2}LD^{-1/2} = I - D^{-1/2}AD^{-1/2}$;
- the *PageRank Laplacian* $\mathcal{L}_{PR} = I - D^{-1}A$.

Remark A.6. Note that $D^{-1/2}$ and D^{-1} are not well defined if there is an isolated node (a node of degree 0). We can either assume there is no such node in our graph, or by convention we let $D_{ii}^{-1/2} = D_{ii}^{-1} = 0$ if i isolated.

Lemma A.6. *If i and j are two neighboring nodes, we express this as $i \sim j$. Furthermore, assume the graph does not have self-loops. Then, we have:*

$$L_{ij} = \begin{cases} d_i & \text{if } i = j, \\ -1 & \text{if } i \sim j, \\ 0 & \text{otherwise} \end{cases} \quad \text{and} \quad \mathcal{L}_{ij} = \begin{cases} 1 & \text{if } i = j, \\ -\dfrac{1}{\sqrt{d_i d_j}} & \text{if } i \sim j, \\ 0 & \text{otherwise.} \end{cases}$$

Proof. This is direct from the definitions of L and $\mathcal{L}$. □

Basic properties of the Laplacians

Proposition A.10. *The standard Laplacian $L = D - A$ has the following properties:*

1. *For any vector $x \in \mathbb{R}^n$ we have $x^T Lx = \frac{1}{2}\sum_{i=1}^{n}\sum_{j=1}^{n} a_{ij}\left(x_i - x_j\right)^2$. More generally, for any matrix $X \in \mathbb{R}^{n\times K}$ we have*

$$\mathrm{Tr}\left(X^T LX\right) = \frac{1}{2}\sum_{k=1}^{K}\sum_{i,j} a_{ij}\left(X_{ik} - X_{jk}\right)^2;$$

2. *L is symmetric and positive semi-definite;*
3. *L has n non-negative, real valued eigenvalues $0 = \lambda_1 \leq \cdots \leq \lambda_n$. Moreover, $L1_n = 0_n$.*

Proof. 1. Recall that $d_i = \sum_{j=1}^{n} a_{ij}$. We can write

$$\begin{aligned}
\frac{1}{2}\sum_{i,j=1}^{n} a_{ij}\left(x_i - x_j\right)^2 &= \frac{1}{2}\left(\sum_{i,j} a_{ij}x_i^2 - 2\sum_{i,j} a_{ij}x_i x_j + \sum_{i,j} a_{ij}x_j^2\right) \\
&= \frac{1}{2}\left(\sum_{i=1}^{n} d_i x_i^2 - 2\sum_{i,j=1}^{n} x_i x_j a_{ij} + \sum_{j=1}^{n} d_j x_j^2\right) \\
&= \sum_{i=1}^{n} d_i x_i^2 - \sum_{i,j=1}^{n} x_i x_j a_{ij} \\
&= x^T Dx - x^T Ax \\
&= x^T Lx.
\end{aligned}$$

More generally, for $X \in \mathbb{R}^{n\times K}$ we notice that

$$\mathrm{Tr}\left(X^T LX\right) = \sum_{k=1}^{K} X_{.k}^T LX_{.k}$$

where $X_{.k}$ denotes the column k of X, and hence this result then holds by applying the previous result.

2. L is symmetric because D and A are. From point 1, we have $X^T LX \geq 0$, so L is positive semi-definite.
3. L is symmetric, so its eigenvalues are real. It is positive semi-definite, so its eigenvalues are non-negative. Finally, $L1_n = 0_n$ is straightforward using the formula derived in point 1.

□

Proposition A.11. *The normalized Laplacian $\mathcal{L}$ satisfies the following properties:*

1. *For any vector $x \in \mathbb{R}^n$ we have $x^T \mathcal{L} x = \frac{1}{2}\sum_{i=1}^{n}\sum_{j=1}^{n} a_{ij}\left(\frac{x_i}{\sqrt{d_i}} - \frac{x_j}{\sqrt{d_j}}\right)^2$. More generally, for any matrix $X \in \mathbb{R}^{n\times K}$ we have*

$$\operatorname{Tr}\left(X^T \mathcal{L} X\right) = \frac{1}{2}\sum_{k=1}^{K}\sum_{i,j} a_{ij}\left(\frac{X_{ik}}{\sqrt{d_i}} - \frac{X_{jk}}{\sqrt{d_j}}\right)^2;$$

2. *$\mathcal{L}$ is symmetric and positive semi-definite;*
3. *$\mathcal{L}$ has n non-negative, real valued eigenvalues $0 = \lambda_1 \leq \cdots \leq \lambda_n \leq 2$. More, $D^{1/2}1_n$ is an eigenvector of $\mathcal{L}$ associated to the eigenvalue 0.*

The proof of Proposition A.11 is similar to that of Proposition A.10.

Standard Laplacian and the number of connected components

Definition A.13 (Indicator vector of a set)**.** Let U be a subset of the node set V. We define 1_U as the $n \times 1$ vector such that $(1_U)_i = 1$ if $i \in U$, and 0 if $i \notin U$. We let 1_n be the n-by-1 vector of all ones.

Lemma A.7. *$LX = 0 \Leftrightarrow X$ is constant on each connected component of G.*

Proof. Let $V_1, \ldots, V_K$ be the connected components of G. Assume $LX = 0$. Then $X^T LX = 0$, and from the formula of the previous proposition it follows that $\forall i, j \in V_k$: $x_i = x_j$. We conclude that $LX = 0$ implies that X is constant on each connected component of G.

Reciprocally, we can see from the direct computation that if X is constant on each V_k, then $LX = 0$. □

Proposition A.12 (Number of connected components)**.** *Let G be an undirected graph with non-negative weights. Then, the multiplicity k of the eigenvalue 0 of L is equal to the number of connected components $V_1, \ldots, V_k$. Moreover, the eigenspace of eigenvalue 0 (Ker L) is spanned by the indicator vectors $1_{V_1}, \ldots, 1_{V_k}$.*

Proof. If $k = 1$, it means the only eigenvector of 0 is $X = 1_n$, and the graph is connected.

Now suppose $k > 1$. We can assume that the vertices are ordered according to the connected components they belong to. Thus, $L = \operatorname{diag}(L_1, \ldots, L_k)$, where L_i is the Laplacian of the $i - th$ connected component. Each L_i has eigenvalue 0 with multiplicity 1, and the corresponding eigenvector is the constant vector of ones. Thus, $L1_{A_i} = L_i 1_n = 0$, and each 1_{1_i} is eigenvector of L associated to 0. □

Example A.4. Suppose that the graph is connected. Then there is only one connected component ($k = 1$), so $\dim \operatorname{Ker} L = 1$, and the corresponding eigenspace is spanned by 1_n.

A.3 Linear Algebra

A.3.1 Symmetric Matrices

Theorem A.8 (Spectral theorem). *If M is symmetric and real valued, then there exists an orthonormal basis consisting of eigenvectors of M. Moreover, the eigenvalues of M are real.*

Counterexample A.9 (if M has complex entries). $M = \begin{pmatrix} 1 & i \\ i & -1 \end{pmatrix}$ is symmetric but not diagonalizable. Indeed, from a direct computation of its characteristic polynomial, we can see that the only eigenvalue is 0.

Definition A.14. A symmetric matrix M is said to be positive semidefinite (PSD) (resp., positive definite PD) if $\forall x \in \mathbb{R}^n : x^T Mx \geq 0$ (resp., $x^T Mx > 0$).

Example A.5. For all $M \in \mathbb{R}^{n\times n}$, the matrix $M^T M$ is symmetric definite positive.

Lemma A.10. *Let M be a symmetric matrix, and $\lambda_1, \ldots, \lambda_n$ its (real) eigenvalues. M is positive semidefinite (resp., positive definite) iff $\lambda_i \geq 0$ (resp., $\lambda_i > 0$).*

A.3.2 Norms

Definition A.15. Let E be a vector space. A function $N : E \to \mathbb{R}$ is a norm if it satisfies the following properties:

1. (positivity) $\forall x \in E : N(x) \geq 0$;
2. (definiteness) $N(x) = 0 \Rightarrow x = 0_E$;
3. (homogeneity) $\forall x \in E, t \in \mathbb{R} : N(tx) \leq |t|N(x)$;
4. (triangle inequality) $\forall x, y \in E : N(x + y) \leq N(x) + N(y)$.

Vector norms

Proposition A.13. *Let $E = \mathbb{R}^n$ and $p \geq 1$, we define the ℓ^p-norms as follows:*

$$\|x\|_p = \left(\sum_{i=1}^{n} |x_i|^p \right)^{1/p}.$$

Proof. It is straightforward to show that $\|.\|_p$ verify the first three conditions. The triangle inequality holds thanks to Minkowski inequality. □

Example A.6. Let $x = (x_1, \ldots, x_n)^T \in \mathbb{R}^n$. We have the following particular cases of ℓ^p-norms:

1. for $p = 1$, $\|x\|_1 = \sum_{i=1}^{n} |x_i|$;
2. for $p = 2$, $\|x\|_2 = \sqrt{\sum_{i=1}^{n} |x_i|^2}$ is the Euclidian norm;
3. for $p = \infty$, we define $\|x\|_\infty = \lim_{p \to +\infty} \|x\|_p = \max\{|x_1|, \ldots, |x_n|\}$.

Moreover, if we introduce the scalar product $< x, y >= x^T y$ between two vectors $x, y \in \mathbb{R}^n$, then $\|x\|_2^2 = x^T x$.

Matrix norms (Serre, 2010)

Definition A.16. Let $\|.\|$ be a norm on $\mathbb{R}^n$, we define the operator norm $|||.|||$ on $\mathbb{R}^{n \times n}$ induced by $\|.\|$ as

$$|||A||| = \sup_{x \in \mathbb{R}^n :\, x \neq 0_n} \frac{\|Ax\|}{\|x\|}.$$

By abuse of notation, we often denote by $\|.\|$ the operator norm instead of $|||.|||$.

Lemma A.11. *Let $A \in \mathbb{R}^{n \times n}$.*

$$|||A||| = \sup_{\|x\|=1} \|Ax\| = \sup_{\|x\| \leq 1} \|Ax\| = \max_{\|x\| \leq 1} \|Ax\|.$$

Example A.7. Let $A \in \mathbb{R}^{n \times n}$. We have the followig induced norms:

1. $|||A|||_1 = \sup_{\|x\|_1 = 1} \|Ax\|_1 = \max_{j=1 \ldots n} \sum_{i=1}^{n} |A_{ij}|$ (max column-sum);
2. $|||A|||_\infty = \sup_{\|x\|_\infty = 1} \|Ax\|_\infty = \max_{i=1 \ldots n} \sum_{j=1}^{n} |A_{ij}|$ (max row-sum);
3. $|||A|||_2 = \sup_{x^T x = 1} \sqrt{x^T A^T A x} = \sqrt{\lambda_{\max}(A^T A)}$, where $\lambda_{\max}(A^T A)$ denotes the largest eigenvalue of the (symmetric) matrix $A^T A$;
4. if A is invertible, then $|||A^{-1}|||_2 = \dfrac{1}{\lambda_{\min}\left(A^T A\right)}$, where $\lambda_{\min}\left(A^T A\right)$ is the smallest eigenvalue of $A^T A$ (non-zero if A^{-1} is invertible).

Proposition A.14. *Let $|||\cdot|||$ be an induced operator norm. Then $\forall A, B \in \mathbb{R}^{n \times n}$: $|||AB||| \leq |||A||| \times |||B|||$.*

Counterexample A.12. This inequality is false in general if the norm is not induced from a vector norm. For example, let $N(A) = \max_{i,j} |a_{ij}|$ (not to be confused with $|||\cdot|||_\infty$), and $A = \begin{pmatrix} 1 & 1 \\ 1 & 1 \end{pmatrix}$. Then, $N(A^2) = 2 > N(A)N(A) = 1$.

Definition A.17. We denote $\|A\|_F = \sqrt{\sum_{i=1}^{n}\sum_{j=1}^{n} |A_{ij}|^2}$ the Frobenius (or Hilbert–Schmidt) norm of a matrix $A \in \mathbb{R}^{n\times n}$.

A.3.3 Courant-Fisher Theorem

Theorem A.13 (Courant-Fisher theorem). *Let $M \in \mathbb{R}^{n\times n}$ be a n-by-n symmetric matrix, and $\lambda_1 \le \cdots \le \lambda_n$ the eigenvalues of A, with associated normalized eigenvectors $v_1, \ldots, v_n$. We have*

$$\lambda_1 = \min_{x\in\mathbb{R}^n:\ \|x\|=1} x^T Mx = \min_{x\in\mathbb{R}^n:\ x\neq 0_n} \frac{x^T Mx}{x^T x}, \tag{A.1}$$

$$\lambda_2 = \min_{\substack{x\in\mathbb{R}^n \\ \|x\|=1 \\ x\perp v_1}} x^T Mx = \min_{\substack{x\in\mathbb{R}^n \\ x\neq 0_n \\ x\perp v_1}} \frac{x^T Mx}{x^T x}, \tag{A.2}$$

$$\lambda_n = \max_{\substack{x\in\mathbb{R}^n \\ \|x\|=1}} x^T Mx = \max_{\substack{x\in\mathbb{R}^n \\ x\neq 0_n}} \frac{x^T Mx}{x^T x}. \tag{A.3}$$

Moreover, the respective arg min *are obtained by v_1, v_2 and v_n, respectively.*

Proof. Let us give two proofs, one by diagonalizing the matrix M, and one using calculus (Lagrange minimisers).

(i) First proof. M being symmetric, we can write $M = P^T DP$. Let $y = Px$. Note that $\|y\| = \|x\|$, thus the constraint $\|x\| = 1$ becomes $\sum_{i=1}^n y_i^2 = 1$. Since $x^T Mx = y^T Dy = \sum_{i=1}^n \lambda_i y_i^2$, this expression is minimised (given the constraint) when all y_i are zeros except for $y_1 = 1$, and maximised when $y_n = 1$ and all others y_i are 0. If $x \perp v_1$, then $y_1 = 0$ is further imposed, and $y^T Dy$ is minimised if $y_2 = 1$ and other y_i are null.

(ii) Second proof. The Lagrangian associated to the minimisation problem (A.1) (or (A.3)) is $\mathcal{L}(x, \lambda) = x^T Mx - \lambda\left(x^T x - 1\right)$. Note that letting $\frac{\partial \mathcal{L}}{\partial \lambda} = 0$ gives back the constraint $\|x\| = 1$. Moreover, $\frac{\partial L}{\partial x} = 2Mx - 2\lambda x$, an hence $\frac{\partial L}{\partial x} = 0$ leads to $Mx = \lambda x$. Thus, x is an eigenvector of M and λ is the corresponding eigenvalue. As Equation (A.1) is a minimisation problem, its solution is the smallest eigenvalue. Similarly, the solution of equation (A.3) is the largest eigenvalue. Finally, the solution is the second smallest eigenvalue if we further impose that $x \perp v_1$. □

Proposition A.15. *Let $M \in \mathbb{R}^{n \times n}$ be a symmetric matrix with $v_1, \dots, v_n$ being an orthonormal basis of eigenvectors associated with $\lambda_1 \leq \lambda_2 \leq \cdots \leq \lambda_n$. The solution of the optimisation problem*

$$\underset{\substack{H \in \mathbb{R}^{n \times K} \\ H^T H = I_K}}{\arg\min} \operatorname{Tr}\left(H^T M H\right)$$

is given by $H = [v_1, \dots, v_K]$.

Proof. Consider the Lagrangian $\mathcal{L}(H, \Lambda) = \operatorname{Tr}\left(H^T M H\right) - \operatorname{Tr}\Big(\Lambda^T (H^T H - I_K)\Big)$, where $\Lambda \in \mathbb{R}^{K \times K}$ is a diagonal matrix, whose entries are the Lagrange multipliers. Since $\frac{\partial \mathcal{L}}{\partial H} = 2MH - 2H\Lambda$, the condition $\frac{\partial \mathcal{L}}{\partial H} = 0$ leads to $MH = H\Lambda$. Thus, the columns of H are indeed eigenvectors of M, and the diagonals elements of Φ are the corresponding eigenvalues. □

A.4 Calculus on Graphs

We refer the reader to (Hein *et al.*, 2007) for additional details on the topic of this section.

A.4.1 Basic Reminders

Consider a function $f \colon \mathbb{R}^n \to \mathbb{R}$. The *gradient* of f at a point $x \in \mathbb{R}^n$ is the vector $\nabla f(x) = \operatorname{grad} f(x) = \left(\frac{\partial f}{\partial x_1}(x), \cdots, \frac{\partial f}{\partial x_n}(x)\right)^T$. The *divergence* is defined for every $x \in \mathbb{R}^n$ as $\operatorname{div} f(x) = \sum_{i=1}^n \frac{\partial f}{\partial x_i}(x)$, and the *Laplacian operator* is $\Delta f(x) = \sum_{i=1}^n \frac{\partial^2 f}{\partial x_i^2}(x)$. In particular, we have $\operatorname{div}\left(\operatorname{grad} f\right)(x) = \Delta f$.

A.4.2 Extension on Graphs

In this section, we consider a directed and weighted graph $G = (V, E)$ whose weights are w_{ij} and node set is $V = \{1, \dots, n\}$.

Functions on graph

We denote by $\mathcal{F}(V)$ the set of *node functions* $f \colon V \to \mathbb{R}$. Since $|V| = n$, any node function f can be represented as a n-by-1 vector $\left(f(1), \dots, f(n)\right)^T$ and $\mathcal{F}(V) \cong$

$\mathbb{R}^n$. In particular, $\mathcal{F}(V)$ is a n-dimensional Hilbert-space whose inner product is

$$< f, g >_{\mathcal{F}(V)} = \sum_{v_i \in V} f(i) g(i)$$

and associated norm $\|f\|_{\mathcal{F}(V)} = \sqrt{< f, f >_{\mathcal{F}(V)}}$.

Similarly, the space of *edge functions* is $\mathcal{F}(E) = \{F \colon E \to \mathbb{R}\}$. This space is equivalent to $\mathbb{R}^{|E|}$, and we introduce the inner product

$$< F, G >_{\mathcal{F}(E)} = \sum_{(i,j) \in E} F(i,j) G(i,j),$$

and the norm $\|F\|_{\mathcal{F}(E)} = \sqrt{< F, F >_{\mathcal{F}(E)}}$. Finally, we can trivially extend any edge function $F \colon E \to \mathbb{R}$ to a function $\tilde{F} \colon V \times V \to \mathbb{R}$ by letting $\tilde{F}(v_i, v_j) = 0$ if $(v_i, v_j) \notin E$. By a slight abuse of notation, we still denote by F the extended function.

Differential graphs operators

Let $\gamma : \mathbb{R}_+ \to \mathbb{R}_+$ such that $\gamma(0) = 0$. The choice of γ will be discussed later. The *graph derivative* of f along a directed edge $(i,j) \in E$ is

$$\frac{\partial f}{\partial j}(i) = \gamma(w_{ij}) \left(f(j) - f(i) \right),$$

and denoted $\partial_j f(i)$ for convenience. In particular, $\partial_i f(i) = 0$ and $f(i) = f(j)$ implies $\partial_j f(i) = 0$.

The *graph gradient* of a node function $f \in \mathcal{F}(V)$ is denoted grad$f$, and is defined by

$$\forall (i,j) \in E: \qquad (\operatorname{grad} f)(i,j) = \partial_j f(i).$$

Hence grad: $\mathcal{F}(V) \to \mathcal{F}(E)$ is a linear operator. The *graph divergence* div is defined to be the adjoint operator[1] of grad, that is

$$< \operatorname{grad} f, G >_{\mathcal{F}(E)} = < f, \operatorname{div} G >_{\mathcal{F}(V)}, \quad \forall f \in \mathcal{F}(V), \forall G \in \mathcal{F}(E).$$

Lemma A.14. *The divergence* div: $\mathcal{F}(E) \to \mathcal{F}(V)$ *of the gradient operator is given by:*

$$(\operatorname{div} G)(i) = \sum_j \gamma(w_{ji}) G(j,i) - \gamma(w_{ij}) G(i,j).$$

1. The adjoint is well defined here since the considered Hilbert spaces have finite dimensions.

Proof. We have

$$
\begin{aligned}
< \operatorname{grad} f, G >_{\mathcal{F}(E)} &= \sum_{i,j} \gamma\left(w_{ij}\right)\left(f(j) - f(i)\right) G(i,j) \\
&= \sum_{i,j} \gamma\left(w_{ij}\right) f(j) G(i,j) - \sum_{i,j} \gamma\left(w_{ij}\right) f(i) G(i,j) \\
&= \sum_{i,j} \gamma\left(w_{ji}\right) f(i) G(j,i) - \sum_{i,j} \gamma\left(w_{ij}\right) f(i) G(i,j) \\
&= \sum_{i} f(i)\left(\gamma\left(w_{ji}\right) G(j,i) - \gamma\left(w_{ij}\right) G(i,j)\right) \\
&= < f, \operatorname{div} G >_{\mathcal{F}(V)} .
\end{aligned}
$$

□

For undirected graphs ($w_{ij} = w_{ji}$), the divergence reduces to

$$
(\operatorname{div} G)(i) = \sum_{j} \gamma\left(w_{ij}\right)\left(G(j,i) - G(i,j)\right).
$$

Finally, we define the *graph Laplacian* $\Delta_\gamma : \mathcal{H}(V) \rightarrow \mathcal{H}(V)$ such that $\Delta_\gamma f = \operatorname{div}\left(\operatorname{grad} f\right)$ for every $f \in \mathcal{F}(V)$.

Lemma A.15. *Let $f \in \mathcal{F}(V)$ and $i \in V$. We have:*

$$
\left(\Delta_\gamma f\right)(i) = \sum_{j}\left(\gamma\left(w_{ij}\right)^2 + \gamma\left(w_{ji}\right)^2\right)\left(f(i) - f(j)\right).
$$

Proof. Let $\gamma_{ij} = \gamma\left(w_{ij}\right)$. We have

$$
\begin{aligned}
\operatorname{div}\left(\operatorname{grad} f\right)(i) &= \sum_{j} \gamma_{ji} \operatorname{grad} f(j,i) - \gamma_{ij} \operatorname{grad} f(i,j) \\
&= \sum_{j} \gamma_{ji}^2 \left(f(i) - f(j)\right) - \gamma_{ij}^2 \left(f(j) - f(i)\right) \\
&= \sum_{j} \left(\gamma_{ij}^2 + \gamma_{ji}^2\right)\left(f(i) - f(j)\right).
\end{aligned}
$$

□

For an undirected graph, the Laplacian operator reduces to

$$
\Delta_\gamma f(i) = 2 \sum_{j} \gamma\left(w_{ij}\right)^2 \left(f(i) - f(j)\right).
$$

Recall the standard Laplacian $L = D - A$, and let $f : V \to \mathbb{R}$ be a node function represented by a n-by-1 vector. We have

$$(Lf)_i = d_i f_i - \sum_j w_{ij} f_j = \sum_j w_{ij} (f_i - f_j) .$$

Thus, $L = \Delta_\gamma$ if $\gamma(x) = \sqrt{x}$. Similarly, the random walk Laplacian $L_{rw} = I - D^{-1}A$ verifies

$$(L_{rw} f)_i = f_i - \sum_j \frac{w_{ij}}{d_i} f_j = \sum_j \frac{w_{ij}}{d_i} (f_i - f_j) ,$$

where the second equality holds since $\sum_j \frac{w_{ij}}{d_i} = 1$. Hence, $L_{rw} = \Delta_\gamma$, if γ verifies $\gamma(w_{ij}) = \sqrt{\frac{w_{ij}}{d_i}}$ for all i, j.

DOI: 10.1561/9781638280514.ch9

Appendix B

Additional Lemmas Related to the Proof of Theorem 5.5

B.1 Mean-field Solution of the Secular Equation (5.19)

B.1.1 Spectral Study of a Perturbed Rank-2 Matrix

Lemma B.1 (Matrix determinant lemma). Suppose $A \in \mathbb{R}^{n\times n}$ is invertible, and let U, V be two n by m matrices. Then,

$$\det(A + UV^T) = \det A \det(I_m + V^T A^{-1} U).$$

Proof. We take the determinant of

$$\begin{pmatrix} A & -U \\ V^T & I \end{pmatrix} = \begin{pmatrix} A & 0 \\ V^T & I \end{pmatrix} \begin{pmatrix} I & -A^{-1}U \\ 0 & I + V^T A^{-1} U \end{pmatrix},$$

and we note that $\det \begin{pmatrix} A & -U \\ V^T & I \end{pmatrix} = \det I \det \left(A + UV^T\right)$ by the Schur complement formula (Horn and Johnson, 2012, Section 0.8.5). $\square$

Proposition B.1. *Let* $M = ZBZ^T$, *where* $B = \begin{pmatrix} a & b \\ b & a \end{pmatrix}$ *is a* 2×2 *matrix, and* $Z = \begin{pmatrix} 1_{n/2} & 0_{n/2} \\ 0_{n/2} & 1_{n/2} \end{pmatrix}$ *is an* $n \times 2$ *matrix. Let* m *be an even number. We denote by* $P_{\mathcal{L}}$ *the* $n \times n$ *diagonal matrix, whose first* $\frac{m}{2}$ *and last* $\frac{m}{2}$ *diagonal elements are ones, all other elements being zeros. Then,*

$$\det\left(tI_n + \lambda P_{\mathcal{L}} - M\right) = t^{n-m-2}(t+\lambda)^{m-2}(t-t_1^+)(t-t_1^-)(t-t_2^+)(t-t_2^-),$$

with

$$t_1^{\pm} = \frac{1}{2}\left(\frac{n}{2}(a+b) - \lambda \pm \sqrt{\left(\lambda + \frac{n}{2}(a+b)\right)^2 - 2(a+b)\lambda m}\right),$$

$$t_2^{\pm} = \frac{1}{2}\left(\frac{n}{2}(a-b) - \lambda \pm \sqrt{\left(\lambda + \frac{n}{2}(a-b)\right)^2 - 2(a-b)\lambda m}\right).$$

Proof. For now, assume that $t \neq -\lambda$ and $t \neq 0$. Then, $tI_n + \lambda P_{\mathcal{L}}$ is invertible, and by Lemma B.1,

$$\begin{aligned}\det\left(tI_n + \lambda P_{\mathcal{L}} - M\right) &= \det(tI_n + \lambda P_{\mathcal{L}}) \det\left(I_2 + Z^T(tI_n + \lambda P_{\mathcal{L}})^{-1}(-ZB)\right) \\ &= (t+\lambda)^m t^{n-m} \det\left(I_2 - Z^T(tI_n + \lambda P_{\mathcal{L}})^{-1} ZB\right).\end{aligned} \tag{B.1}$$

Moreover,

$$\left(tI_n + \lambda P_{\mathcal{L}}\right)^{-1} = \frac{1}{t}(I_n - P_{\mathcal{L}}) + \frac{1}{t+\lambda} P_{\mathcal{L}} = \frac{1}{t} I_n - \frac{\lambda}{t(t+\lambda)} P_{\mathcal{L}}.$$

Therefore, we can write

$$\begin{aligned} Z^T\left(tI_n + \lambda P_{\mathcal{L}}\right)^{-1} ZB &= \frac{1}{t} Z^T ZB - \frac{\lambda}{t(t+\lambda)} Z^T P_{\mathcal{L}} ZB \\ &= \frac{1}{t}\frac{n}{2} B - \frac{\lambda}{t(t+\lambda)} \frac{m}{2} B = xB, \end{aligned}$$

where $x := \frac{n}{2} \frac{1}{t(t+\lambda)} \left(t + \lambda\left(1 - \frac{m}{n}\right)\right)$. Thus, a direct computation of the determinant gives

$$\det\left(I_2 - Z^T\left(tI_n + \lambda P_{\mathcal{L}}\right)^{-1} ZB\right) = \left(1 - x(a+b)\right)\left(1 - x(a-b)\right).$$

Going back to equation (B.1), we can write

$$\det\left(tI_n + \lambda P_{\mathcal{L}} - M\right) = (t+\lambda)^{m-2} t^{n-m-2} P_1(t) P_2(t), \tag{B.2}$$

with $P_1(t) = t(t+\lambda) - \frac{n}{2}(a+b)\left(t + \lambda(1-\frac{m}{n})\right)$ and $P_2(t) = t(t+\lambda) - \frac{n}{2}(a - b)\left(t + \lambda(1-\frac{m}{n})\right)$. Since $t \in \mathbb{R} \mapsto \det(tI_n + \lambda P_{\mathcal{L}} - M)$ is continuous (even analytic), expression (B.2) is also valid for $t = 0$ and $t = -\lambda$ (Avrachenkov *et al.*, 2013a). We end the proof by observing that

$$P_1(t) = (t - t_1^+)(t - t_1^-) \qquad \text{and} \qquad P_2(t) = (t - t_2^+)(t - t_2^-),$$

where $t_1^\pm$ and $t_2^\pm$ are defined in the proposition's statement. □

Corollary B.2. *Let A be the adjacency matrix of a DC-SBM with $p_{\text{in}} > p_{\text{out}} > 0$, and s be the oracle information. Let $\lambda, \tau > 0$, and $\bar{d}_\tau = \frac{n}{2}\left(p_{\text{in}} + p_{\text{out}}\right) - n\tau$, $\bar{\alpha} = \frac{n}{2}\left(p_{\text{in}} - p_{\text{out}}\right)$. Let $A_\tau := A - \tau 1_n 1_n^T$ and $P_{\mathcal{L}}$ be the diagonal matrix whose element $(P_{\mathcal{L}})_{ii}$ is 1 if $s_i \neq 0$, and 0 otherwise. Then, the spectrum of $\mathbb{E}\tilde{\mathcal{L}} = -\mathbb{E}A_\tau + \lambda \mathcal{P} - \gamma I_n$ is $\left\{-\gamma - t_1^\pm; -\gamma - t_2^\pm; -\gamma; -\gamma + \lambda; 0\right\}$, where*

$$t_1^\pm = \frac{1}{2}\left(\bar{d}_\tau - \lambda \pm \sqrt{\left(\lambda + \bar{d}_\tau\right)^2 - 4\bar{d}_\tau \lambda\left(\eta_1 + \eta_0\right)}\right),$$

$$t_2^\pm = \frac{1}{2}\left(\bar{\alpha} - \lambda \pm \sqrt{\left(\lambda + \bar{\alpha}\right)^2 - 4\bar{\alpha}\lambda\left(\eta_1 + \eta_0\right)}\right).$$

Proof. Let $M = \begin{pmatrix} p_{\text{in}} - \tau & p_{\text{out}} - \tau \\ p_{\text{out}} - \tau & p_{\text{in}} - \tau \end{pmatrix}$ and $Z = \begin{pmatrix} 1_{n/2} & 0_{n/2} \\ 0_{n/2} & 1_{n/2} \end{pmatrix}$. Then, we notice that $\mathbb{E}A_\tau = ZMZ^T$ and we can apply Proposition B.1 to compute the characteristic polynomial of $\mathbb{E}\tilde{\mathcal{L}}$. For $x \in \mathbb{R}$, $\det\left(\mathbb{E}\tilde{\mathcal{L}} - xI_n\right) = \det\left((-\gamma - x)I_n - \mathbb{E}A_\tau + \lambda\mathcal{P}\right)$, whose roots are $-\gamma - t_1^\pm, -\gamma - t_2^\pm, -\gamma$, and $-\gamma + \lambda$. □

B.1.2 Estimation of $\bar{\gamma}_*$

Lemma B.3. *Let $\bar{\gamma}_*$ be the solution of equation* (5.19) *for the mean-field model. Then,*

$$-\bar{\alpha}(1 - 2\eta_0) \leq \bar{\gamma}_* \leq -\bar{\alpha}.$$

Proof. For $\lambda \geq 0$, we denote by $(\bar{x}_\lambda, \bar{\gamma}_*(\lambda))$ the solution of the system (5.17) on a mean-field DC-SBM. The proof is in two steps. First, let us show that $\bar{\gamma}_*(0) = -\bar{\alpha}$ and $\bar{\gamma}_*(\infty) = -\bar{\alpha}(1 - 2\eta_0)$. For $\lambda = 0$, the constrained linear system (5.17) reduces to an eigenvector problem, and hence $\bar{\gamma}_*(0)$ equals $-\alpha$, the smallest eigenvalue of $-\mathbb{E}A_\tau$. Moreover, when $\lambda = \infty$, the hard constraint $x_\ell = \bar{s}_\ell$ is enforced, and the system (5.17) becomes

$$\begin{cases} (-\mathbb{E}A_\tau - \bar{\gamma}_*(\infty)I_n)_{uu}\bar{x}_u &= (\mathbb{E}A_\tau)_{u\ell}\bar{s}_\ell \\ \bar{x}_u^T\bar{x}_u &= n(1-\eta_0-\eta_1) \end{cases}$$

and we verify by hand that $\bar{\gamma}_*(\infty) = -\bar{\alpha}(1-2\eta_0)$ together with $\bar{x}_u = Z_u$ is indeed the solution.

Second, if we let $C_\lambda(x) = -x^T\mathbb{E}A_\tau x + \lambda(\bar{s}-\mathcal{P}x)^T(\bar{s}-\mathcal{P}x)$ be the cost function minimized in (5.14), then from equation (5.17) we have $\bar{\gamma}_*(\lambda_1) - \bar{\gamma}_*(\lambda_2) = C_{\lambda_1}(\bar{x}_1) - C_{\lambda_2}(\bar{x}_2) + \lambda_1\bar{x}_1^T\bar{s} - \lambda_2\bar{x}_2^T\bar{s}$. Since $\lambda \mapsto C_\lambda(x)$ is increasing, then $\lambda_1 \leq \lambda_2$ implies $C_{\lambda_1}(\bar{x}_1) \leq C_{\lambda_2}(\bar{x}_2)$. Since $\bar{x}_\lambda^T\bar{s} \geq 0$ (if it was not the case, then $C_\lambda(-\bar{x}_\lambda) \leq C_\lambda(\bar{x}_\lambda)$, and hence $\bar{x}_\lambda \neq \arg\min_{x\in\mathbb{R}^n} C_\lambda(x)$), we can conclude that $\bar{\gamma}_*(0) \leq \bar{\gamma}_*(\lambda)$ and that $\bar{\gamma}_*(\lambda) \leq \bar{\gamma}_*(\infty)$. □

B.1.3 Concentration of γ_*

Proposition B.2. *Let γ_* and $\bar{\gamma}_*$ be the solutions of equation* (5.17) *for a DC-SBM and the mean-field DC-SBM, respectively. Then*

$$|\gamma_* - \bar{\gamma}_*| \leq \left(1 + \frac{27\,(\bar{\alpha}+\lambda)^3}{\sqrt{2}\sqrt{\eta_1+\eta_0}(\eta_1-\eta_0)\bar{\alpha}^2\lambda}\right)\sqrt{\bar{d}}.$$

Proof. The gradient with respect to $(\bar{\delta}_1, ..., \bar{\delta}_n, \bar{b}_1, ..., \bar{b}_n, \gamma)$ of the left-hand-side of equation (5.19) is equal to

$$2\sum_{i=1}^{n}\frac{\bar{b}_i}{\bar{\delta}_i-\bar{\gamma}}\left[\frac{\Delta b_i}{\bar{\delta}_i-\bar{\gamma}_*} - \frac{\bar{b}_i\Delta\delta_i}{(\bar{\delta}_i-\bar{\gamma}_*)^2} + \frac{\bar{b}_i\Delta\gamma}{(\bar{\delta}_i-\bar{\gamma}_*)^2}\right].$$

Thus, we have

$$\Delta\gamma\sum_{i=1}^{n}\frac{\bar{b}_i^2}{(\bar{\delta}_i-\bar{\gamma}_*)^3} = \sum_{i=1}^{n}\frac{\bar{b}_i^2}{(\bar{\delta}_i-\bar{\gamma}_*)^3}\Delta\delta_i - \sum_{i=1}^{n}\frac{\bar{b}_i}{(\bar{\delta}_i-\bar{\gamma}_*)^2}\Delta b_i + o\left(\Delta\delta_i, \Delta b_i\right).$$

Firstly, we see that for all $i \in [n]$, $\Delta\delta_i = \left|\delta_i - \bar{\delta}_i\right| \leq \|A - \mathbb{E}A\| \leq \bar{d}$ by the concentration of the adjacency matrix of a DC-SBM graph. Therefore, using this fact and $\bar{\gamma}_* \leq \bar{\delta}_1 \leq \bar{\delta}_2 \leq \cdots \leq \bar{\delta}_n$,

$$\begin{aligned}\Delta\gamma = |\gamma_* - \bar{\gamma}_*| &\leq \max_i\left|\delta_i-\bar{\delta}_i\right| + \frac{\max_i\frac{1}{(\bar{\delta}_i-\bar{\gamma}_*)^2}}{\min_i\frac{1}{(\bar{\delta}_i-\bar{\gamma}_*)^3}}\frac{\sum_i|\bar{b}_i|\cdot|b_i-\bar{b}_i|}{\sum_i\bar{b}_i^2} \\ &\leq \sqrt{\bar{d}} + \frac{\max_i\left(\bar{\delta}_i-\bar{\gamma}_*\right)^3}{\min_i\left(\bar{\delta}_i-\bar{\gamma}_*\right)^2}\frac{\sum_i|\bar{b}_i|\cdot|b_i-\bar{b}_i|}{\sum_i\bar{b}_i^2}.\end{aligned}$$

We notice that $\min_i |\bar{\delta}_i - \bar{\gamma}_*| = \bar{\delta}_1 - \bar{\gamma}_*$. By using Lemma B.3 and the expression of $\bar{\delta}_1$ given in Corollary B.2, we have

$$\min_i |\bar{\delta}_i - \bar{\gamma}_*| \;\geq\; \bar{\alpha} + \lambda.$$

Similarly, $\max_i |\bar{\delta}_i - \bar{\gamma}_*| = \bar{\delta}_n - \bar{\gamma}_* = \bar{\delta}_n - \bar{\delta}_1 + \bar{\delta}_1 - \bar{\gamma}_*$. Corollary B.2 implies $\bar{\delta}_n = \lambda$ and $\bar{\delta}_1 = \frac{1}{2}\left(\lambda - \bar{\alpha} - \sqrt{(\lambda + \bar{\alpha})^2 - 4\bar{\alpha}\lambda(\eta_0 + \eta_1)}\right)$, thus $\bar{\delta}_n - \bar{\delta}_1 \leq \bar{\alpha} + \lambda$. Hence, using Lemma B.3,

$$\max_i |\bar{\delta}_i - \bar{\gamma}_*| \;\leq\; \frac{3}{2}\left(\bar{\alpha} + \lambda\right).$$

Therefore, we have

$$|\gamma_* - \bar{\gamma}_*| \;\leq\; \sqrt{\bar{d}} + \frac{27}{8}(\bar{\alpha} + \lambda) \cdot \frac{\sum_i |\bar{b}_i| \cdot |b_i - \bar{b}_i|}{\sum_i \bar{b}_i^2}. \tag{B.3}$$

The term $\frac{\sum_i |\bar{b}_i| \cdot |b_i - \bar{b}_i|}{\sum_i \bar{b}_i^2}$ can be bounded as follow. Let $\mathcal{I} = \{i \in [n] \colon \bar{b}_i \neq 0\}$. Then

$$\sum_i |\bar{b}_i| \cdot |b_i - \bar{b}_i| \;\leq\; \max_{i \in \mathcal{I}} |b_i - \bar{b}_i| \cdot \sum_{i \in \mathcal{I}} |\bar{b}_i|.$$

Combining the Cauchy-Schwarz inequality

$$\left|b_i - \bar{b}_i\right| \;=\; \lambda \left|(Q_{\cdot i} - \bar{Q}_{\cdot i})^T \bar{s}\right| \;\leq\; \lambda \left\|Q_{\cdot i} - \bar{Q}_{\cdot i}\right\|_2 \cdot \|\bar{s}\|,$$

with the Davis-Kahan theorem (Yu *et al.*, 2015)

$$\left\|Q_{\cdot i} - \bar{Q}_{\cdot i}\right\|_2 \;\leq\; \frac{2^{3/2}\,\|A - \mathbb{E}A\|}{\min\left\{\bar{\delta}_i - \bar{\delta}_{i-1}, \bar{\delta}_{i+1} - \bar{\delta}_i\right\}},$$

$\|\bar{s}\| = \sqrt{(\eta_0 + \eta_1)n}$, and the concentration of A towards $\mathbb{E}A$, yields

$$\max_{i \in \mathcal{I}} |b_i - \bar{b}_i| \;\leq\; \frac{\lambda\sqrt{(\eta_0 + \eta_1)n}}{\min_{i \in \mathcal{I}}\left\{\bar{\delta}_i - \bar{\delta}_{i-1}, \bar{\delta}_{i+1} - \bar{\delta}_i\right\}} \cdot 2^{3/2}\sqrt{\bar{d}}.$$

Using Lemma B.4, we see that $\mathcal{I} = \{i \in [n] : \delta_i \notin \{0, t_1^-\}\}$. Combining it with Corollary B.2, gives

$$\begin{aligned}\min_{i \in \mathcal{I}} \left\{\bar{\delta}_i - \bar{\delta}_{i-1}, \bar{\delta}_{i+1} - \bar{\delta}_i\right\} &= \lambda + t_2^+ \\ &= \frac{\alpha + \lambda}{2}\left(1 - \sqrt{1 - 4\frac{\alpha\lambda}{(\alpha+\lambda)^2}(\eta_0 + \eta_1)}\right) \\ &\geq \frac{\alpha\lambda}{\alpha + \lambda}(\eta_0 + \eta_1),\end{aligned}$$

where we used $\sqrt{1-x} \leq 1 - x/2$. Therefore,

$$\sum_i \left|\bar{b}_i\right| \cdot \left|b_i - \bar{b}_i\right| \leq 2^{3/2}\sqrt{\frac{n\bar{d}}{\eta_0 + \eta_1}} \cdot \frac{\alpha + \lambda}{\alpha} \cdot \sum_i \left|\bar{b}_i\right|.$$

By noticing that $\sum_i \bar{b}_i^2 \geq \left(\sum_i \left|\bar{b}_i\right|\right)^2 \geq \left|\bar{b}_1\right| \cdot \sum_i \left|\bar{b}_i\right| \geq \sqrt{n}\frac{\eta_1 - \eta_0}{2}\frac{\bar{\alpha}\lambda}{\lambda + \bar{\alpha}}\sum_i \left|\bar{b}_i\right|$ where we used $\bar{b}_1 \geq \sqrt{n}\frac{\eta_1 - \eta_0}{2}\frac{\bar{\alpha}\lambda}{\lambda + \bar{\alpha}}$ (Lemma B.4), we have

$$\frac{\sum_i \left|\bar{b}_i\right| \cdot \left|b_i - \bar{b}_i\right|}{\sum_i \bar{b}_i^2} \leq \frac{2^{5/2}}{(\eta_1 - \eta_0)\sqrt{\eta_1 + \eta_0}}\frac{(\alpha + \lambda)^2}{\alpha^2\lambda}\sqrt{\bar{d}}.$$

Going back to inequality (B.3), this implies that

$$|\gamma_* - \bar{\gamma}_*| \leq \left(1 + \frac{27\,(\bar{\alpha} + \lambda)^3}{\sqrt{2}\sqrt{\eta_1 + \eta_0}(\eta_1 - \eta_0)\bar{\alpha}^2\lambda}\right)\sqrt{\bar{d}}.$$

□

Lemma B.4. *Let* $-\mathbb{E}A\tau + \lambda\mathcal{P} = \bar{Q}\bar{\Delta}\bar{Q}^T$, *where* $\bar{\Delta} = \operatorname{diag}\left(\bar{\delta}_1, \ldots, \bar{\delta}_n\right)$ *and* $\bar{Q}^T\bar{Q} = I_n$. *Denote* $\bar{b} = \lambda\bar{Q}^T s$. *We have* $\bar{b}_1 \geq \sqrt{n}\frac{\lambda(\eta_1 - \eta_0)}{2}\frac{\bar{\alpha}}{\lambda + \bar{\alpha}}$. *Moreover,* $\bar{b}_i = 0$ *if* $\bar{\delta}_i = 0$ *or if* $\bar{\delta}_i = -t_1^-$.

Proof. First, from Corollary B.2,

$$\bar{\delta}_1 = -t_2^+ = -\frac{1}{2}\left(\bar{\alpha} - \lambda + \sqrt{\left(\lambda + \bar{\alpha}\right)^2 - 4\bar{\alpha}\lambda\,(\eta_1 + \eta_0)}\right).$$

By symmetry, the i-th component of the first eigenvector $\bar{Q}_{\cdot 1}$ (associated with $\bar{\delta}_1$) is equal to

$$\begin{cases} v_1 Z_i & \text{if } i \in [\ell], \\ v_0 Z_i & \text{if } i \notin [\ell], \end{cases}$$

where v_1 and v_0 are to be determined. Thus, the equation $(-\mathbb{E}A_\tau + \lambda\mathcal{P})\,\bar{Q}_{\cdot 1} = \bar{\delta}_1 \bar{Q}_{\cdot 1}$ leads to

$$\begin{cases} \bar{\alpha}\left((\eta_1+\eta_0)v_1 + (1-\eta_1-\eta_0)v_0\right) &= -t_2^+ v_0 \\ \bar{\alpha}\left((\eta_1+\eta_0)v_1 + (1-\eta_1-\eta_0)v_0\right) + \lambda v_1 &= -t_2^+ v_1, \end{cases}$$

which, given the norm constraint $\|v\|_2 = 1$, yields

$$\begin{cases} v_1 &= \dfrac{1}{\sqrt{n}} \dfrac{t_2^+}{\sqrt{(\eta_1+\eta_0)\left(t_2^+\right)^2 + (1-\eta_1-\eta_0)\left(t_2^+ + \lambda\right)^2}}, \\ v_0 &= \dfrac{1}{\sqrt{n}} \dfrac{+t_2^+ + \lambda}{\sqrt{(\eta_1+\eta_0)\left(t_2^+\right)^2 + (1-\eta_1-\eta_0)\left(t_2^+ + \lambda\right)^2}}. \end{cases}$$

Since $\bar{b}_1 = \lambda v^T \bar{s} = \lambda(\eta_1 - \eta_0) n v_1$, we have

$$\frac{\bar{b}_1}{\sqrt{n}} = \lambda(\eta_1-\eta_0) \frac{t_2^+}{\sqrt{(\eta_1+\eta_0)\left(t_2^+\right)^2 + (1-\eta_1-\eta_0)\left(t_2^+ + \lambda\right)^2}}.$$

The proof ends by noticing that $t_2^+ \geq \frac{\bar{\alpha}}{2}$ and $t_2^+ \leq \bar{\alpha}$. Indeed,

$$\begin{aligned} \frac{\bar{b}_1}{\sqrt{n}} &\geq \lambda(\eta_1-\eta_0) \frac{\bar{\alpha}}{2\sqrt{(\eta_1+\eta_0)\bar{\alpha}^2 + (1-\eta_1-\eta_0)(\bar{\alpha}+\lambda)^2}} \\ &\geq \frac{\lambda(\eta_1-\eta_0)}{2} \frac{\bar{\alpha}}{(\bar{\alpha}+\lambda)\sqrt{(\eta_1+\eta_0)\left(\frac{\bar{\alpha}}{\bar{\alpha}+\lambda}\right)^2 + 1 - \eta_1 - \eta_0}} \\ &\geq \frac{\lambda(\eta_1-\eta_0)}{2} \frac{\bar{\alpha}}{\lambda+\bar{\alpha}}. \end{aligned}$$

This proves the first claim of the lemma.

Similarly, by symmetry the i-th component of the eigenvector v' associated with $-t_1^-$ equals v'_ℓ if $i \in \ell$, and v'_u otherwise, and therefore $(v')^T s = 0$.

Finally, let $I_0 := \{i \in [n] : \bar{\delta}_i = 0\}$. By Corollary B.2, we have $|I_0| = n(1 - \eta_1 - \eta_0) - 2$. Since 0 is also eigenvalue of order $n(1-\eta_0-\eta_1)-2$ of the extracted sub-matrix $(-\mathbb{E}A_\tau + \lambda\mathcal{P})_{u,u} = (-\mathbb{E}A_\tau)_{u,u}$, we have for all $k \in I_0$, $\bar{Q}_{ik} = 0$ for every $i \in [n]$. Hence, for $k \in I_0$, $b_k = \lambda \bar{Q}_{\cdot k}^T s = 0$. □

B.2 Mean-field Solution of the Constrained Linear System (5.17)

In this section, we calculate the solution $\bar{x}$ to the mean-field model and deduce from it the conditions to recover the clusters.

Proposition B.3. *Suppose that $\tau > p_{\text{out}}$. Then, the solution of equation (5.18) on the mean-field DC-SBM is the vector $\bar{x}$, whose element $\bar{x}_i$ is given by*

$$\bar{x}_i = \begin{cases} C\left(-1 + (\eta_1 - \eta_0)\bar{\alpha}B\right) Z_i, & \text{if } i \in \ell \text{ and } s_i \neq Z_i, \\ C\left(1 + (\eta_1 - \eta_0)\bar{\alpha}B\right) Z_i, & \text{if } i \in \ell \text{ and } s_i = Z_i, \\ \frac{-\bar{\alpha}C}{\bar{\alpha}(1-\eta_1-\eta_0)+\bar{\gamma}_*}(\eta_1 - \eta_0)\left(1 + (\eta_1 + \eta_0)\bar{\alpha}B\right) Z_i, & \text{if } i \notin \ell, \end{cases}$$

where $\bar{\alpha} = \frac{n}{2}(p_{\text{in}} - p_{\text{out}})$, $B = \frac{\bar{\alpha}\bar{\gamma}_*}{\lambda\bar{\alpha}(1-\eta_1-\eta_0)+\bar{\gamma}_*(\lambda-\bar{\alpha}-\bar{\gamma}_*)}$ *and* $C = \frac{\lambda}{\lambda-\bar{\gamma}_*}$.

Proof. Let $\bar{x}$ be a solution of equation (5.18). By symmetry, we have

$$\bar{x}_i = \begin{cases} x_t Z_i, & \text{if } i \in [\ell] \text{ and } \bar{s}_i = Z_i, \\ x_f Z_i, & \text{if } i \in [\ell] \text{ and } \bar{s}_i = -Z_i, \\ x_0 Z_i, & \text{if } i \notin [\ell], \end{cases}$$

where x_t, x_f and x_0 are unknowns to be determined. Since for every $i \in [n]$,

$$(\mathbb{E}A_\tau \bar{x})_i = \bar{\alpha}\left(x_0(1 - \eta_1 - \eta_0) + x_t\eta_1 + x_f\eta_0\right),$$

the linear system composed of the equations $\left((-\mathbb{E}A_\tau + \lambda\mathcal{P} - \bar{\gamma}_* I_n)\bar{x}\right)_i = \lambda s_i$ for all $i \in [n]$ leads to the system

$$\begin{cases} -\bar{\alpha}\left((1 - \eta_1 - \eta_0)x_0 + x_t\eta_1 + x_f\eta_0\right) - \bar{\gamma}_* x_0 = 0, \\ -\bar{\alpha}\left((1 - \eta_1 - \eta_0)x_0 + x_t\eta_1 + x_f\eta_0\right) - \bar{\gamma}_* x_t + \lambda x_t = \lambda, \\ -\bar{\alpha}\left((1 - \eta_1 - \eta_0)x_0 + x_t\eta_1 + x_f\eta_0\right) - \bar{\gamma}_* x_f + \lambda x_f = -\lambda. \end{cases}$$

The rows of the latter system correspond to a node unlabeled by the oracle, correctly labeled and falsely labeled, respectively. This system can be rewritten as follows:

$$\begin{cases} x_0 = \frac{-\bar{\alpha}}{\bar{\alpha}(1-\eta_1-\eta_0)+\bar{\gamma}_*}\left(\eta_1 x_t + \eta_0 x_f\right), \\ \bar{\gamma}_* x_0 + x_t(\lambda - \bar{\gamma}_*) = \lambda, \\ \bar{\gamma}_* x_0 + x_f(\lambda - \bar{\gamma}_*) = -\lambda. \end{cases}$$

In particular, we have $x_t - x_f = \frac{2\lambda}{\lambda-\bar{\gamma}_*}$. By subsequently eliminating x_0 and x_t in the equation $\bar{\gamma}_* x_0 + x_f(\lambda - \bar{\gamma}_*) = -\lambda$, we find

$$x_f = \frac{\lambda}{\lambda - \bar{\gamma}_*}\left(-1 + \frac{\bar{\alpha}\bar{\gamma}_*(\eta_1 - \eta_0)}{\lambda\bar{\alpha}(1 - \eta_1 - \eta_0) + \lambda\bar{\gamma}_* - \bar{\gamma}_*(\bar{\alpha} + \bar{\gamma}_*)}\right),$$

$$x_t = \frac{\lambda}{\lambda - \bar{\gamma}_*}\left(1 + \frac{\bar{\alpha}\bar{\gamma}_*(\eta_1 - \eta_0)}{\lambda\bar{\alpha}(1 - \eta_1 - \eta_0) + \lambda\bar{\gamma}_* - \bar{\gamma}_*(\bar{\alpha} + \bar{\gamma}_*)}\right),$$

and finally

$$x_0 = \frac{-\bar{\alpha}}{\bar{\alpha}(1-\eta_1-\eta_0)+\bar{\gamma}_*} \cdot \frac{\lambda}{\lambda-\bar{\gamma}_*}\left(1+\frac{\bar{\alpha}\bar{\gamma}_*(\eta_1+\eta_0)}{\lambda\bar{\alpha}(1-\eta_1-\eta_0)+\lambda\bar{\gamma}_*-\bar{\gamma}_*(\bar{\alpha}+\bar{\gamma}_*)}\right).$$

□

Corollary B.5. *Suppose that* $\tau > p_{\text{out}}$. *Then* $\text{sign}(\bar{x}_i) = \text{sign}(Z_i)$ *if*

- *node* i *is not labeled by the oracle;*
- *node* i *is correctly labeled by the oracle;*
- *node* i *is mislabeled by the oracle and* $\lambda < (1-2\eta_0)\bar{\alpha}\frac{\eta_1-\eta_0}{\eta_1+\eta_0}$.

Proof. A node i is correctly classified by decision rule (5.16) if the sign of $\bar{x}_i$ is equal to the sign of Z_i. Using Lemma B.3 in Appendix B.1.2, we have $-\bar{\alpha} \le \bar{\gamma}_* \le -\bar{\alpha}(1-2\eta_0)$. Therefore, the quantities B and C in Proposition B.3 verify $C \ge 0$ and $\frac{1-2\eta_0}{\lambda(\eta_0+\eta_1)} \le B \le \frac{1}{\lambda(\eta_0+\eta_1)}$. The statement then follows from the expression of $\bar{x}_i$ computed in Proposition B.3. □

References

Abbe, E. 2018. "Community detection and stochastic block models". *Foundations and Trends in Communications and Information Theory*. 14(1–2), 1–162.

Abbe, E., A. S. Bandeira, and G. Hall. 2015. "Exact recovery in the stochastic block model". *IEEE Transactions on Information Theory*. 62(1): 471–487.

Abbe, E., J. Fan, K. Wang, Y. Zhong, *et al.* 2020. "Entrywise eigenvector analysis of random matrices with low expected rank". *Annals of Statistics*. 48(3): 1452–1474.

Adamic, L. A. and N. Glance. 2005. "The political blogosphere and the 2004 US election: Divided they blog". In: *Proceedings of the 3rd International Workshop on Link Discovery*. 36–43.

Agirre, E. and A. Soroa. 2009. "Personalizing pagerank for word sense disambiguation". In: *Proceedings of the 12th Conference of the European Chapter of the ACL (EACL 2009)*. 33–41.

Akgün, M. K. and M. K. Tural. 2020. "k-step betweenness centrality". *Computational and Mathematical Organization Theory*. 26(1): 55–87.

Albert, R., H. Jeong, and A.-L. Barabási. 2000. "Error and attack tolerance of complex networks". *Nature*. 406(6794): 378–382.

Aldous, D. and J. Fill. 2002. *Reversible Markov Chains and Random Walks on Graphs*. Berkeley. URL: http://www.stat.berkeley.edu/~aldous/RWG/book.html.

Altman, A. and M. Tennenholtz. 2005. "Ranking systems: The PageRank axioms". In: *Proceedings of the 6th ACM Conference on Electronic Commerce*. 1–8.

Amini, A. A., E. Levina, *et al.* 2018. "On semidefinite relaxations for the block model". *Annals of Statistics*. 46(1): 149–179.

Andersen, R., F. Chung, and K. Lang. 2006. "Local graph partitioning using pagerank vectors". In: *2006 47th Annual IEEE Symposium on Foundations of Computer Science (FOCS'06)*. IEEE. 475–486.

Arenas, A., A. Fernandez, and S. Gomez. 2008. "Analysis of the structure of complex networks at different resolution levels". *New Journal of Physics*. 10(5): 053039.

Avrachenkov, K. E., J. A. Filar, and P. G. Howlett. 2013a. *Analytic Perturbation Theory and its Applications*. SIAM.

Avrachenkov, K. E., A. Y. Kondratev, V. V. Mazalov, and D. G. Rubanov. 2018a. "Network partitioning algorithms as cooperative games". *Computational Social Networks*. 5(1): 1–28.

Avrachenkov, K. E., V. V. Mazalov, and B. T. Tsynguev. 2015. "Beta current flow centrality for weighted networks". In: *International Conference on Computational Social Networks*. Springer. 216–227.

Avrachenkov, K., A. Bobu, and M. Dreveton. 2021a. "Higher-order spectral clustering for geometric graphs". *Journal of Fourier Analysis and Applications*. 27(2): 1–29.

Avrachenkov, K., V. S. Borkar, A. Kadavankandy, and J. K. Sreedharan. 2018b. "Revisiting random walk based sampling in networks: Evasion of burn-in period and frequent regenerations". *Computational Social Networks*. 5(1): 1–19.

Avrachenkov, K., V. S. Borkar, and K. Saboo. 2016a. "Distributed and asynchronous methods for semi-supervised learning". In: *International Workshop on Algorithms and Models for the Web-Graph*. 34–46.

Avrachenkov, K., P. Chebotarev, and D. Rubanov. 2019. "Similarities on graphs: Kernels versus proximity measures". *European Journal of Combinatorics*. 80: 47–56.

Avrachenkov, K., V. Dobrynin, D. Nemirovsky, S. K. Pham, and E. Smirnova. 2008a. "Pagerank based clustering of hypertext document collections". In: *Proceedings of the 31st ACM SIGIR*. 873–874.

Avrachenkov, K. and M. Dreveton. 2019. "Almost exact recovery in label spreading". In: *International Workshop on Algorithms and Models for the Web-Graph*. 30–43.

Avrachenkov, K. and M. Dreveton. 2020. "Almost exact recovery in noisy semi-supervised learning". *arXiv preprint arXiv:2007.14717*.

Avrachenkov, K., M. Dreveton, and L. Leskelä. 2021b. "Recovering communities in temporal networks using persistent edges". In: *International Conference on Computational Data and Social Networks*. Springer. 243–254.

Avrachenkov, K., R. v. d. Hofstad, and M. Sokol. 2014a. "Personalized pagerank with node-dependent restart". In: *International Workshop on Algorithms and Models for the Web-Graph*. 23–33.

Avrachenkov, K., A. Kadavankandy, and N. Litvak. 2018c. "Mean field analysis of personalized PageRank with implications for local graph clustering". *Journal of Statistical Physics*. 173(3–4): 895–916.

Avrachenkov, K., L. Leskelä, and M. Dreveton. 2022. "Community recovery in non-binary and temporal stochastic block models". *arXiv preprint arXiv:2008.04790.*

Avrachenkov, K., N. Litvak, V. Medyanikov, and M. Sokol. 2013b. "Alpha current flow betweenness centrality". In: *International Workshop on Algorithms and Models for the Web-Graph.* Springer. 106–117.

Avrachenkov, K., N. Litvak, D. Nemirovsky, E. Smirnova, and M. Sokol. 2011. "Quick detection of top-k personalized pagerank lists". In: *International Workshop on Algorithms and Models for the Web-Graph.* Springer. 50–61.

Avrachenkov, K., N. Litvak, and K. S. Pham. 2008b. "A singular perturbation approach for choosing the PageRank damping factor". *Internet Mathematics.* 5(1–2): 47–69.

Avrachenkov, K., N. Litvak, L. O. Prokhorenkova, and E. Suyargulova. 2014b. "Quick detection of high-degree entities in large directed networks". In: *2014 IEEE International Conference on Data Mining.* IEEE. 20–29.

Avrachenkov, K., N. Litvak, M. Sokol, and D. Towsley. 2014c. "Quick detection of nodes with large degrees". *Internet Mathematics.* 10(1–2): 1–19.

Avrachenkov, K., A. Mishenin, P. Gonçalves, and M. Sokol. 2012. "Generalized optimization framework for graph-based semi-supervised learning". In: *Proceedings of the 2012 SIAM International Conference on Data Mining.* SIAM. 966–974.

Avrachenkov, K., G. Neglia, and A. Tuholukova. 2016b. "Subsampling for chain-referral methods". In: *International Conference on Analytical and Stochastic Modeling Techniques and Applications.* Springer. 17–31.

Avrachenkov, K., A. Piunovskiy, and Y. Zhang. 2018d. "Hitting times in Markov chains with restart and their application to network centrality". *Methodology and Computing in Applied Probability.* 20(4): 1173–1188.

Avrachenkov, K., B. Ribeiro, and J. K. Sreedharan. 2016c. "Inference in OSNs via lightweight partial crawls". *Proceedings of ACM SIGMETRICS.* 44(1): 165–177.

Avrachenkov, K., B. Ribeiro, and D. Towsley. 2010. "Improving random walk estimation accuracy with uniform restarts". In: *International Workshop on Algorithms and Models for the Web-Graph (WAW).* Springer. 98–109.

Barabási, A.-L. 2016. *Network Science.* Cambridge University Press.

Barabási, A.-L. and R. Albert. 1999. "Emergence of scaling in random networks". *Science.* 286(5439): 509–512.

Barucca, P., F. Lillo, P. Mazzarisi, and D. Tantari. 2018. "Disentangling group and link persistence in dynamic stochastic block models". *Journal of Statistical Mechanics: Theory and Experiment.* 2018(12): 123407.

Bastian, M., S. Heymann, and M. Jacomy. 2009. "Gephi: An open source software for exploring and manipulating networks". In: *Proceedings of the International AAAI Conference on Web and Social Media*, Vol. 3. No. 1. 361–362.

Batagelj, V. and U. Brandes. 2005. "Efficient generation of large random networks". *Physical Review E*. 71(3): 036113.

Bavelas, A. 1950. "Communication patterns in task-oriented groups". *Journal of the Acoustical Society of America*. 22(6): 725–730.

Belkin, M. and P. Niyogi. 2002. "Using manifold structure for partially labelled classification". In: *Proceedings of the 15th International Conference on Neural Information Processing Systems*. Cambridge, MA, USA: MIT Press. 953–960.

Ben-David, S., T. Lu, and D. Pál. 2008. "Does unlabeled data provably help? Worst-case analysis of the sample complexity of semi-supervised learning". In: *Proceedings of Conference on Learning Theory*.

Bergstrom, C. 2007. "Eigenfactor: Measuring the value and prestige of scholarly journals". *College & Research Libraries News*. 68(5): 314–316.

Bergstrom, C. T., J. D. West, and M. A. Wiseman. 2008. "The EigenfactorTM metrics". *Journal of Neuroscience*. 28(45): 11433–11434.

Bhattacharyya, S. and S. Chatterjee. 2020. "General community detection with optimal recovery conditions for multi-relational sparse networks with dependent layers". *arXiv preprint arXiv:2004.03480*.

Bickel, P. J. and P. Sarkar. 2016. "Hypothesis testing for automated community detection in networks". *Journal of the Royal Statistical Society: Series B (Statistical Methodology)*. 78(1): 253–273.

Billingsley, P. 1961. "Statistical methods in Markov chains". *Annals of Mathematical Statistics*. 32(1): 12–40.

Blondel, V. D., J.-L. Guillaume, R. Lambiotte, and E. Lefebvre. 2008. "Fast unfolding of communities in large networks". *Journal of Statistical Mechanics: Theory and Experiment*. 2008(10): P10008.

Bojchevski, A., J. Klicpera, B. Perozzi, A. Kapoor, M. Blais, B. Rózemberczki, M. Lukasik, and S. Günnemann. 2020. "Scaling Graph Neural Networks with approximate PageRank". In: *Proceedings of the 26th ACM SIGKDD International Conference on Knowledge Discovery & Data Mining*. 2464–2473.

Boldi, P., F. Bonchi, C. Castillo, D. Donato, A. Gionis, and S. Vigna. 2008. "The query-flow graph: model and applications". In: *Proceedings of the 17th ACM Conference on Information and Knowledge Management*. 609–618.

Boldi, P. and S. Vigna. 2014. "Axioms for centrality". *Internet Mathematics*. 10(3–4): 222–262.

Bollen, J., M. A. Rodriquez, and H. Van de Sompel. 2006. "Journal status". *Scientometrics*. 69(3): 669–687.

Bollobás, B. 2001. *Random Graphs*. No. 73. Cambridge University Press.

Bollobás, B., O. Riordan, J. Spencer, and G. Tusnády. 2001. "The degree sequence of a scale-free random graph process". *Random Structures & Algorithms*. 18(3): 279–290.

Bonacich, P. 1987. "Power and centrality: A family of measures". *American Journal of Sociology*. 92(5): 1170–1182.

Bonacich, P. and P. Lloyd. 2001. "Eigenvector-like measures of centrality for asymmetric relations". *Social Networks*. 23(3): 191–201.

Borassi, M. and E. Natale. 2019. "KADABRA is an adaptive algorithm for betweenness via random approximation". *Journal of Experimental Algorithmics (JEA)*. 24: 1–35.

Borgatti, S. P. 2005. "Centrality and network flow". *Social Networks*. 27(1): 55–71.

Borgatti, S. P., M. G. Everett, and J. C. Johnson. 2018. *Analyzing Social Networks*. 2nd ed. SAGE.

Brandes, U. 2008. "On variants of shortest-path betweenness centrality and their generic computation". *Social Networks*. 30(2): 136–145.

Brandes, U., D. Delling, M. Gaertler, R. Gorke, M. Hoefer, Z. Nikoloski, and D. Wagner. 2007. "On modularity clustering". *IEEE Transactions on Knowledge and Data Engineering*. 20(2): 172–188.

Brandes, U. and D. Fleischer. 2005. "Centrality measures based on current flow". In: *Annual Symposium on Theoretical Aspects of Computer Science*. Springer. 533–544.

Brauer, A. 1952. "Limits for the characteristic roots of a matrix. IV: Applications to stochastic matrices". *Duke Mathematical Journal*. 19(1): 75–91.

Brémaud, P. 1999. *Markov chains: Gibbs fields, Monte Carlo simulation, and queues*, Vol. 31. Springer.

Brin, S. and L. Page. 1998. "The anatomy of a large-scale hypertextual web search engine". *Computer Networks and ISDN Systems*. 30(1–7): 107–117.

Broido, A. D. and A. Clauset. 2019. "Scale-free networks are rare". *Nature Communications*. 10(1): 1–10.

Calder, J., B. Cook, M. Thorpe, and D. Slepcev. 2020. "Poisson learning: Graph based semi-supervised learning at very low label rates". In: *Proceedings of International Conference on Machine Learning (ICML 2020)*. PMLR. 1306–1316.

Callaghan, T., P. J. Mucha, and M. A. Porter. 2007. "Random walker ranking for NCAA division IA football". *The American Mathematical Monthly*. 114(9): 761–777.

Carley, K. M. and D. Skillicorn. 2005. "Special issue on analyzing large scale networks: The Enron corpus". *Computational & Mathematical Organization Theory*. 11(3): 179–181.

Carrington, P. J., J. Scott, and S. Wasserman. 2005. *Models and Methods in Social Network Analysis*. Cambridge University Press.

Chandra, A. K., P. Raghavan, W. L. Ruzzo, R. Smolensky, and P. Tiwari. 1996. "The electrical resistance of a graph captures its commute and cover times". *Computational Complexity*. 6(4): 312–340.

Chapelle, O., B. Schölkopf, and A. Zien. 2006. *Semi-Supervised Learning*. Adaptive Computation and Machine Learning. MIT Press.

Chen, M., Z. Wei, B. Ding, Y. Li, Y. Yuan, X. Du, and J.-R. Wen. 2020. "Scalable graph neural networks via bidirectional propagation". *Advances in Neural Information Processing Systems*. 33: 14556–14566.

Chen, P., H. Xie, S. Maslov, and S. Redner. 2007. "Finding scientific gems with Google's PageRank algorithm". *Journal of Informetrics*. 1(1): 8–15.

Chen, Y., Y. Chi, J. Fan, C. Ma, *et al.* 2021. "Spectral methods for data science: A statistical perspective". *Foundations and Trends in Machine Learning*. 14(5): 566–806.

Cherven, K. 2015. *Mastering Gephi Network Visualization*. Packt Publishing.

Chien, E., J. Peng, P. Li, and O. Milenkovic. 2020. "Adaptive universal generalized PageRank graph neural network". *arXiv preprint arXiv:2006.07988*.

Chung, F. 2007. "The heat kernel as the pagerank of a graph". *Proceedings of the National Academy of Sciences*. 104(50): 19735–19740.

Chung, F. and L. Lu. 2006. *Complex Graphs and Networks (CBMS Regional Conference Series in Mathematics)*. Boston, MA, USA: American Mathematical Society. ISBN: 0821836579.

Clauset, A., M. E. Newman, and C. Moore. 2004. "Finding community structure in very large networks". *Physical Review E*. 70(6): 066111.

Clauset, A., C. R. Shalizi, and M. E. Newman. 2009. "Power-law distributions in empirical data". *SIAM Review*. 51(4): 661–703.

Clemente, G. P. and A. Cornaro. 2020. "A novel measure of edge and vertex centrality for assessing robustness in complex networks". *Soft Computing*. 24(18): 13687–13704.

Cohen, E. and H. Kaplan. 2007. "Spatially-decaying aggregation over a network". *Journal of Computer and System Sciences*. 73(3): 265–288.

Cooper, C., T. Radzik, and Y. Siantos. 2016. "Fast low-cost estimation of network properties using random walks". *Internet Mathematics*. 12(4): 221–238.

Cozman, F. G., I. Cohen, and M. Cirelo. 2002. "Unlabeled data can degrade classification performance of generative classifiers". In: *Proceedings of Flairs-02*. 327–331.

Dasgupta, A., R. Kumar, and D. Sivakumar. 2012. "Social sampling". In: *Proceedings of the 18th ACM SIGKDD*. 235–243.

Davoodi, E., K. Kianmehr, and M. Afsharchi. 2013. "A semantic social network-based expert recommender system". *Applied Intelligence*. 39(1): 1–13.

De Nooy, W., A. Mrvar, and V. Batagelj. 2018. *Exploratory Social Network Analysis with Pajek: Revised and expanded edition for updated software*. 3rd ed. Cambridge University Press.

Decelle, A., F. Krzakala, C. Moore, and L. Zdeborová. 2011. "Asymptotic analysis of the stochastic block model for modular networks and its algorithmic applications". *Physical Review E*. 84(6): 066106.

Defferrard, M., X. Bresson, and P. Vandergheynst. 2016. "Convolutional neural networks on graphs with fast localized spectral filtering". *Advances in Neural Information Processing Systems*. 29.

Dekker, A. 2005. "Conceptual distance in social network analysis". *Journal of Social Structure*. 6(3): 31.

Demmel, J. W., O. A. Marques, B. N. Parlett, and C. Vömel. 2008. "Performance and accuracy of LAPACK's symmetric tridiagonal eigensolvers". *SIAM Journal on Scientific Computing*. 30(3): 1508–1526.

Dempster, A. P., N. M. Laird, and D. B. Rubin. 1977. "Maximum likelihood from incomplete data via the EM algorithm". *Journal of the Royal Statistical Society: Series B (Methodological)*. 39(1): 1–22.

Dhara, S., J. Gaudio, E. Mossel, and C. Sandon. 2022. "Spectral recovery of binary censored block models". In: *Proceedings of the 2022 Annual ACM-SIAM Symposium on Discrete Algorithms (SODA)*. SIAM. 3389–3416.

Ding, Y., E. Yan, A. Frazho, and J. Caverlee. 2009. "PageRank for ranking authors in co-citation networks". *Journal of the American Society for Information Science and Technology*. 60(11): 2229–2243.

Dodds, P. S., R. Muhamad, and D. J. Watts. 2003. "An experimental study of search in global social networks". *Science*. 301(5634): 827–829.

Doreian, P., V. Batagelj, and A. Ferligoj. 2005. *Generalized Blockmodeling*. Cambridge University Press.

Draief, M. and L. Massoulié. 2010. *Epidemics and Rumours in Complex Networks*. Cambridge University Press.

Durrett, R. 2007. *Random Graph Dynamics*, Vol. 200. Cambridge University Press.

Ellens, W., F. M. Spieksma, P. Van Mieghem, A. Jamakovic, and R. E. Kooij. 2011. "Effective graph resistance". *Linear Algebra and its Applications*. 435(10): 2491–2506.

Ellson, J., E. R. Gansner, E. Koutsofios, S. C. North, and G. Woodhull. 2004. "Graphviz and dynagraph–static and dynamic graph drawing tools". In: *Graph Drawing Software*. Springer. 127–148.

Erdős, P. and A. Rényi. 1959. "On random graphs". *Publicationes Mathematicae, Debrecen*. 6: 290–297.

Erickson, B. H. 1979. "Some problems of inference from chain data". *Sociological Methodology*. 10: 276–302.

Estrada, E. and N. Hatano. 2008. "Communicability in complex networks". *Physical Review E.* 77(3): 036111.

Estrada, E. and J. A. Rodriguez-Velazquez. 2005. "Subgraph centrality in complex networks". *Physical Review E.* 71(5): 056103.

Everett, M. G. and S. P. Borgatti. 1999. "The centrality of groups and classes". *Journal of Mathematical Sociology.* 23(3): 181–201.

Fan, C., L. Zeng, Y. Ding, M. Chen, Y. Sun, and Z. Liu. 2019. "Learning to identify high betweenness centrality nodes from scratch: A novel graph neural network approach". In: *Proceedings of the 28th ACM CIKM'19.* 559–568.

Fei, Y. and Y. Chen. 2019. "Achieving the bayes error rate in stochastic block model by sdp, robustly". In: *Conference on Learning Theory.* PMLR. 1235–1269.

Feige, U. and E. Ofek. 2005. "Spectral techniques applied to sparse random graphs". *Random Structures & Algorithms.* 27(2): 251–275.

Fiala, D. 2012. "Time-aware PageRank for bibliographic networks". *Journal of Informetrics.* 6(3): 370–388.

Fiala, D., F. Rousselot, and K. Ježek. 2008. "PageRank for bibliographic networks". *Scientometrics.* 76(1): 135–158.

Fortunato, S. 2010. "Community detection in graphs". *Physics Reports.* 486(3–5): 75–174.

Fortunato, S. and M. Barthelemy. 2007. "Resolution limit in community detection". *Proceedings of the National Academy of Sciences.* 104(1): 36–41.

Fournet, J. and A. Barrat. 2014. "Contact patterns among high school students". *PLOS ONE.* 9(9): 1–17.

Fouss, F., K. Francoisse, L. Yen, A. Pirotte, and M. Saerens. 2012. "An experimental investigation of kernels on graphs for collaborative recommendation and semisupervised classification". *Neural Networks.* 31: 53–72.

Fouss, F., A. Pirotte, J.-M. Renders, and M. Saerens. 2007. "Random-walk computation of similarities between nodes of a graph with application to collaborative recommendation". *IEEE Transactions on knowledge and data engineering.* 19(3): 355–369.

Freeman, L. C. 1977. "A set of measures of centrality based on betweenness". *Sociometry* 35–41.

Freeman, L. C., S. P. Borgatti, and D. R. White. 1991. "Centrality in valued graphs: A measure of betweenness based on network flow". *Social networks.* 13(2): 141–154.

Friedkin, N. E. 1991. "Theoretical foundations for centrality measures". *American Journal of Sociology.* 96(6): 1478–1504.

Galhotra, S., A. Mazumdar, S. Pal, and B. Saha. 2018. "The geometric block model". In: *Thirty-Second AAAI Conference on Artificial Intelligence.*

Gander, W., G. H. Golub, and U. Von Matt. 1989. "A constrained eigenvalue problem". *Linear Algebra and its Applications*. 114: 815–839.

Garey, M. R., D. S. Johnson, and L. Stockmeyer. 1974. "Some simplified NP-complete problems". In: *Proceedings of the 6-th ACM Symposium on Theory of Computing*. ACM. 47–63.

Gauvin, W., B. Ribeiro, D. Towsley, B. Liu, and J. Wang. 2010. "Measurement and gender-specific analysis of user publishing characteristics on myspace". *IEEE Network*. 24(5): 38–43.

Getoor, L. 2005. "Link-based classification". In: *Advanced Methods for Knowledge Discovery from Complex Data*. Springer, 189–207.

Ghasemian, A. 2019. "Limits of model selection, link prediction, and community detection". *PhD thesis*. University of Colorado at Boulder.

Ghasemian, A., P. Zhang, A. Clauset, C. Moore, and L. Peel. 2016. "Detectability thresholds and optimal algorithms for community structure in dynamic networks". *Physical Review X*. 6(3): 031005.

Gilbert, E. N. 1959. "Random graphs". *Annals of Mathematical Statistics*. 30(4): 1141–1144.

Gjoka, M., M. Kurant, C. T. Butts, and A. Markopoulou. 2010. "Walking in facebook: A case study of unbiased sampling of OSNs". In: *Proceedings of IEEE Infocom 2010*. 1–9.

Gleich, D. F. 2015. "PageRank beyond the Web". *SIAM Review*. 57(3): 321–363.

Gleich, D. and M. Mahoney. 2014. "Anti-differentiating approximation algorithms: A case study with min-cuts, spectral, and flow". In: *International Conference on Machine Learning*. PMLR. 1018–1025.

Goldenberg, A., A. X. Zheng, S. E. Fienberg, E. M. Airoldi, *et al.* 2010. "A survey of statistical network models". *Foundations and Trends® in Machine Learning*. 2(2): 129–233.

González-Pereira, B., V. P. Guerrero-Bote, and F. Moya-Anegón. 2010. "A new approach to the metric of journals' scientific prestige: The SJR indicator". *Journal of Informetrics*. 4(3): 379–391.

Good, B. H., Y.-A. De Montjoye, and A. Clauset. 2010. "Performance of modularity maximization in practical contexts". *Physical Review E*. 81(4): 046106.

Goodman, L. A. 1961. "Snowball sampling". *Annals of Mathematical Statistics*: 148–170.

Gori, M., A. Pucci, V. Roma, and I. Siena. 2007. "Itemrank: A random-walk based scoring algorithm for recommender engines". In: *IJCAI*. Vol. 7. 2766–2771.

Grady, L. J. and J. R. Polimeni. 2010. *Discrete Calculus: Applied Analysis on Graphs for Computational Science*. Vol. 3. Springer.

Guédon, O. and R. Vershynin. 2016. "Community detection in sparse networks via Grothendieck's inequality". *Probability Theory and Related Fields*. 165(3): 1025–1049.

Hagberg, A. A., D. A. Schult, and P. J. Swart. 2008. "Exploring Network Structure, Dynamics, and Function using NetworkX". In: G. Varoquaux, T. Vaught, and J. Millman (eds.): *Proceedings of the 7th Python in Science Conference*. Pasadena, Ed. by G. Varoquaux, T. Vaught, and J. Millman. CA USA, 11–15.

Hajek, B., Y. Wu, and J. Xu. 2016a. "Achieving exact cluster recovery threshold via semidefinite programming". *IEEE Transactions on Information Theory*. 62(5): 2788–2797.

Hajek, B., Y. Wu, and J. Xu. 2016b. "Achieving exact cluster recovery threshold via semidefinite programming: Extensions". *IEEE Transactions on Information Theory*. 62(10): 5918–5937.

Hastings, W. K. 1970a. "Monte Carlo Sampling Methods using Markov Chains and their Applications".

Hastings, W. K. 1970b. "Monte Carlo sampling methods using Markov chains and their applications". *Biometrika*. 57: 97–109.

Heckathorn, D. D. 1997. "Respondent-driven sampling: A new approach to the study of hidden populations". *Social Problems*. 44(2): 174–199.

Hein, M., J.-Y. Audibert, and U. v. Luxburg. 2007. "Graph Laplacians and their convergence on random neighborhood graphs". *Journal of Machine Learning Research*. 8(6).

Hofstad, R. van der. 2016. *Random Graphs and Complex Networks*. Vol. 1. *Cambridge Series in Statistical and Probabilistic Mathematics*. Cambridge University Press.

Holland, P. W. and S. Leinhardt. 1981. "An exponential family of probability distributions for directed graphs". *Journal of the American Statistical Association*. 76(373): 33–50.

Holme, P., B. J. Kim, C. N. Yoon, and S. K. Han. 2002. "Attack vulnerability of complex networks". *Physical review E*. 65(5): 056109.

Holme, P. and J. Saramäki. 2012. "Temporal networks". *Physics Reports*. 519(3): 97–125.

Hopcroft, J. and D. Sheldon. 2008. "Manipulation-resistant reputations using hitting time". *Internet Mathematics*. 5(1–2): 71–90.

Horn, R. A. and C. R. Johnson. 2012. *Matrix Analysis*. Cambridge University Press.

Hu, J., H. Qin, T. Yan, and Y. Zhao. 2020. "Corrected Bayesian information criterion for stochastic block models". *Journal of the American Statistical Association*. 115(532): 1771–1783.

Hubbell, C. H. 1965. "An input-output approach to clique identification". *Sociometry* 377–399.

Jackson, M. O. 2010. *Social and Economic Networks*. Princeton University Press.

Jackson, M. O. and A. Wolinsky. 1996. "A strategic model of social and economic networks". *Journal of Economic Theory*. 71(1): 44–74.

Jamonnak, S., J. Kilgallin, C.-C. Chan, and E. Cheng. 2015. "Recommenddit: A Recommendation Service for Reddit Communities". In: *2015 International Conference on Computational Science and Computational Intelligence (CSCI)*. IEEE. 374–379.

Janson, S., T. Luczak, and A. Rucinski. 2011. *Random Graphs*. Vol. 45. John Wiley & Sons.

Jog, V. and P.-L. Loh. 2015. "Recovering communities in weighted stochastic block models". In: *2015 53rd Annual Allerton Conference on Communication, Control, and Computing (Allerton)*. 1308–1315.

Jung, A., A. O. Hero III, A. C. Mara, S. Jahromi, A. Heimowitz, and Y. C. Eldar. 2019. "Semi-supervised learning in network-structured data via total variation minimization". *IEEE Transactions on Signal Processing*. 67(24): 6256–6269.

Karrer, B. and M. E. Newman. 2011. "Stochastic blockmodels and community structure in networks". *Physical Review E*. 83(1): 016107.

Katz, L. 1953. "A new status index derived from sociometric analysis". *Psychometrika*. 18(1): 39–43.

Keener, J. P. 1993. "The Perron–Frobenius theorem and the ranking of football teams". *SIAM Review*. 35(1): 80–93.

Kendall, M. G. 1955. "Further contributions to the theory of paired comparisons". *Biometrics*. 11(1): 43–62.

Kingma, D. P. and M. Welling. 2014. "Auto-Encoding Variational Bayes". In: *Proceedings of the 2nd International Conference on Learning Representations (ICLR)*.

Kipf, T. N. and M. Welling. 2017. "Semi-supervised classification with graph convolutional networks". In: *5th International Conference on Learning Representations*. ICLR.

Kivelä, M., A. Arenas, M. Barthelemy, J. P. Gleeson, Y. Moreno, and M. A. Porter. 2014. "Multilayer networks". *Journal of Complex Networks*. 2(3): 203–271.

Kleinberg, J. M. 1999. "Authoritative sources in a hyperlinked environment". *Journal of ACM*. 46(5): 604–632.

Kleinfeld, J. S. 2002. "The small world problem". *Society*. 39(2): 61–66.

Klicpera, J., A. Bojchevski, and S. Günnemann. 2019. "Predict then Propagate: Graph Neural Networks meet Personalized PageRank". In: *7th International Conference on Learning Representations, ICLR 2019, New Orleans, LA, USA, May 6–9, 2019*.

Knoke, D. and S. Yang. 2019. *Social Network Analysis*. SAGE Publications.

Kolaczyk, E. D., D. B. Chua, and M. Barthélemy. 2009. "Group betweenness and co-betweenness: Inter-related notions of coalition centrality". *Social Networks*. 31(3): 190–203.

Kolaczyk, E. D. and G. Csárdi. 2020. *Statistical Analysis of Network Data with R*. 2nd ed. Springer.

Krioukov, D., F. Papadopoulos, M. Kitsak, A. Vahdat, M. Boguñá. 2010. "Hyperbolic geometry of complex networks". *Physical Review E*. 82(3): 036106.

Krzakala, F., C. Moore, E. Mossel, J. Neeman, A. Sly, L. Zdeborová, and P. Zhang. 2013. "Spectral redemption in clustering sparse networks". *Proceedings of the National Academy of Sciences*. 110(52): 20935–20940.

Kuhn, H. W. 1982. "Nonlinear programming: A historical view". *ACM SIGMAP Bulletin*. (31): 6–18.

Kumar, A., Y. Sabharwal, and S. Sen. 2004. "A simple linear time $(1+\epsilon)$-approximation algorithm for k-means clustering in any dimensions". In: *Proceedings of 45-th IEEE Symposium on Foundations of Computer Science*. IEEE. 454–462.

Landau, E. 1895. "Zur relativen Wertbemessung der Turnierresultate". *Deutsches Wochenschach*. 11(366–369): 3.

Langville, A. N. and C. D. Meyer. 2012. *Who's# 1?: The Science of Rating and Ranking*. Princeton University Press.

Le, C. M. and E. Levina. 2015. "Estimating the number of communities in networks by spectral methods". *arXiv preprint arXiv:1507.00827*.

Le, C. M., E. Levina, and R. Vershynin. 2017. "Concentration and regularization of random graphs". *Random Structures & Algorithms*. 51(3): 538–561.

LeCun, Y., C. Cortes, and C. J. Burges. 1998. *The mnist database of handwritten digits*. URL: http://yann.lecun.com/exdb/mnist/.

Lei, J. 2016. "A goodness-of-fit test for stochastic block models". *Annals of Statistics*. 44(1): 401–424.

Lei, J. and A. Rinaldo. 2015. "Consistency of spectral clustering in stochastic block models". *Annals of Statistics*. 43(1): 215–237.

Lima-Mendez, G. and J. van Helden. 2009. "The powerful law of the power law and other myths in network biology". *Molecular BioSystems*. 5(12): 1482–1493.

Lu, Q. and L. Getoor. 2003. "Link-based classification". In: *Proceedings of the Twentieth International Conference on International Conference on Machine Learning. ICML'03*. Washington, DC, USA: AAAI Press. 496–503.

Lusseau, D., K. Schneider, O. J. Boisseau, P. Haase, E. Slooten, and S. M. Dawson. 2003. "The bottlenose dolphin community of Doubtful Sound features a large proportion of long-lasting associations". *Behavioral Ecology and Sociobiology*. 54(4): 396–405.

Mai, X. and R. Couillet. 2018. "A random matrix analysis and improvement of semi-supervised learning for large dimensional data". *Journal of Machine Learning Research*. 19(1): 3074–3100.

Mai, X. and R. Couillet. 2021. "Consistent Semi-Supervised Graph Regularization for High Dimensional Data". *Journal of Machine Learning Research*. 22(94): 1–48.

Marchiori, M. and V. Latora. 2000. "Harmony in the small-world". *Physica A: Statistical Mechanics and its Applications*. 285(3–4): 539–546.

Mariani, M. S., M. Medo, and Y.-C. Zhang. 2016. "Identification of milestone papers through time-balanced network centrality". *Journal of Informetrics*. 10(4): 1207–1223.

Mastrandrea, R., J. Fournet, and A. Barrat. 2015. "Contact patterns in a high school: A comparison between data collected using wearable sensors, contact diaries and friendship surveys". *PLOS ONE*. 10(9): 1–26.

Masuda, N. and R. Lambiotte. 2021. *A Guide to Temporal Networks*. 2nd ed. World Scientific.

Matias, C. and V. Miele. 2017. "Statistical clustering of temporal networks through a dynamic stochastic block model". *Journal of the Royal Statistical Society: Series B (Statistical Methodology)*. 79(4): 1119–1141.

Mazalov, V. V., K. E. Avrachenkov, L. I. Trukhina, and B. T. Tsynguev. 2016. "Game-theoretic centrality measures for weighted graphs". *Fundamenta Informaticae*. 145(3): 341–358.

Mazalov, V. V. and L. I. Trukhina. 2014. "Generating functions and the Myerson vector in communication networks". *Discrete Mathematics and Applications*. 24(5): 295–303.

Mei, Q., D. Zhou, and K. Church. 2008. "Query suggestion using hitting time". In: *Proceedings of the 17th ACM CIKM'08*. 469–478.

Metropolis, N., A. W. Rosenbluth, M. N. Rosenbluth, A. H. Teller, and E. Teller . 1953. "Equation of state calculations by fast computing machines". *Journal of Chemical Physics*. 21(6): 1087–1092.

Meyer, C. D. 2000. *Matrix Analysis and Applied Linear Algebra*. Vol. 71. SIAM.

Michalak, T. P., K. V. Aadithya, P. L. Szczepanski, B. Ravindran, and N. R. Jennings. 2013. "Efficient computation of the Shapley value for game-theoretic network centrality". *Journal of Artificial Intelligence Research*. 46: 607–650.

Milgram, S. 1967. "The small world problem". *Psychology Today*. 2(1): 60–67.

Moore, C. 2017. "The Computer Science and Physics of Community Detection: Landscapes, Phase Transitions, and Hardness". *Bulletin of EATCS*. 1(121).

Moscato, V., A. Picariello, and G. Sperli. 2019. "Community detection based on game theory". *Engineering Applications of Artificial Intelligence*. 85: 773–782.

Mossel, E., J. Neeman, and A. Sly. 2015. "Consistency thresholds for the planted bisection model". In: *Proceedings of the 47-th ACM Symposium on Theory of Computing*. 69–75.

Mrvar, A. and V. Batagelj. 2016. "Analysis and visualization of large networks with program package Pajek". *Complex Adaptive Systems Modeling*. 4(1): 1–8.

Myerson, R. B. 1977. "Graphs and cooperation in games". *Mathematics of Operations Research*. 2(3): 225–229.

Namata, G., B. London, L. Getoor, and B. Huang. 2012. "Query-driven active surveying for collective classification". In: *10th International Workshop on Mining and Learning with Graphs*. Vol. 8.

Newman, M. 2018. *Networks*. 2nd ed. Oxford University Press,

Newman, M. E. 2001a. "Scientific collaboration networks. I. Network construction and fundamental results". *Physical Review E*. 64(1): 016131.

Newman, M. E. 2001b. "Scientific collaboration networks. II. Shortest paths, weighted networks, and centrality". *Physical Review E*. 64(1): 016132.

Newman, M. E. 2004. "Fast algorithm for detecting community structure in networks". *Physical Review E*. 69(6): 066133.

Newman, M. E. 2005a. "A measure of betweenness centrality based on random walks". *Social Networks*. 27(1): 39–54.

Newman, M. E. 2005b. "Power laws, Pareto distributions and Zipf's law". *Contemporary Physics*. 46(5): 323–351.

Newman, M. E. 2013. "Spectral methods for community detection and graph partitioning". *Physical Review E*. 88(4): 042822.

Newman, M. E. 2016. "Equivalence between modularity optimization and maximum likelihood methods for community detection". *Physical Review E*. 94(5): 052315.

Newman, M. E. and M. Girvan. 2004. "Finding and evaluating community structure in networks". *Physical Review E*. 69(2): 026113.

Ofori-Boateng, D., A. K. Dey, Y. R. Gel, and H. V. Poor. 2021. "Graph-theoretic analysis of power grid robustness". *Advanced Data Analytics for Power Systems*: 175.

Orecchia, L. and Z. A. Zhu. 2014. "Flow-based algorithms for local graph clustering". In: *Proceedings of the 25th ACM-SIAM Symposium on Discrete Algorithms*. 1267–1286.

Orponen, P. and S. E. Schaeffer. 2005. "Local clustering of large graphs by approximate Fiedler vectors". In: *International Workshop on Experimental and Efficient Algorithms*. Springer. 524–533.

Ostuni, V. C., T. Di Noia, E. Di Sciascio, and R. Mirizzi. 2013. "Top-N recommendations from implicit feedback leveraging linked open data". In: *Proceedings of the 7th ACM Conference on Recommender systems*. 85–92.

Pan, R. K. and J. Saramäki. 2011. "Path lengths, correlations, and centrality in temporal networks". *Physical Review E*. 84(1): 016105.

Peixoto, T. P. 2014a. "Efficient Monte Carlo and greedy heuristic for the inference of stochastic block models". *Physical Review E*. 89(1): 012804.

Peixoto, T. P. 2014b. "The graph-tool Python library". URL: https://graph-tool.skewed.de.

Peixoto, T. P. 2019. "Bayesian stochastic blockmodeling". *Advances in Network Clustering and Blockmodeling* 289–332.

Penrose, M. 2003. *Random Geometric Graphs*. Vol. 5. Oxford University Press.

Pinski, G. and F. Narin. 1976. "Citation influence for journal aggregates of scientific publications: Theory, with application to the literature of physics". *Information Processing & Management*. 12(5): 297–312.

Prell, C. 2012. *Social network analysis: History, theory and methodology*. SAGE.

Puterman, M. L. 2014. *Markov decision processes: Discrete stochastic dynamic programming*. John Wiley & Sons.

Ravi, S. and Q. Diao. 2016. "Large scale distributed semi-supervised learning using streaming approximation". In: *Artificial intelligence and statistics*. PMLR. 519–528.

Reichardt, J. and S. Bornholdt. 2006. "Statistical mechanics of community detection". *Physical Review E*. 74(1): 016110.

Ribeiro, B. and D. Towsley. 2010. "Estimating and sampling graphs with multidimensional random walks". In: *Proceedings of the 10th ACM SIGCOMM*. 390–403.

Robert, C. and G. Casella. 2013. *Monte Carlo statistical methods*. Springer.

Rochat, Y. 2009. "Closeness centrality extended to unconnected graphs: The harmonic centrality index". In: *Applications of Social Network Analysis Conference (ASNA)*.

Rosvall, M., D. Axelsson, and C. T. Bergstrom. 2009. "The map equation". *European Physical Journal, Special Topics*. 178(1): 13–23.

Rosvall, M. and C. T. Bergstrom. 2008. "Maps of random walks on complex networks reveal community structure". *Proceedings of the National Academy of Sciences*. 105(4): 1118–1123.

Rueda, D. F., E. Calle, and J. L. Marzo. 2017. "Robustness comparison of 15 real telecommunication networks: Structural and centrality measurements". *Journal of Network and Systems Management*. 25(2): 269–289.

Saade, A., F. Krzakala, and L. Zdeborová. 2014. "Spectral clustering of graphs with the Bethe Hessian". *Advances in Neural Information Processing Systems*. 27.

Sabidussi, G. 1966. "The centrality index of a graph". *Psychometrika*. 31(4): 581–603.

Saldana, D. F., Y. Yu, and Y. Feng. 2017. "How many communities are there?". *Journal of Computational and Graphical Statistics*. 26(1): 171–181.

Salganik, M. J. and D. D. Heckathorn. 2004. "Sampling and estimation in hidden populations using respondent-driven sampling". *Sociological Methodology*. 34(1): 193–240.

Sankararaman, A. and F. Baccelli. 2018. "Community detection on euclidean random graphs". In: *Proceedings of the Twenty-Ninth Annual ACM-SIAM Symposium on Discrete Algorithms*. SIAM. 2181–2200.

Sapiezynski, P., A. Stopczynski, D. D. Lassen, and S. Lehmann. 2019. "Interaction data from the Copenhagen networks study". *Scientific Data*. 6(1): 1–10.

Scarselli, F., M. Gori, A. C. Tsoi, M. Hagenbuchner, and G. Monfardini. 2008. "The graph neural network model". *IEEE transactions on neural networks*. 20(1): 61–80.

Scott, J. and P. J. Carrington. 2011. *The SAGE handbook of social network analysis*. SAGE Publications.

Seeley, J. R. 1949. "The net of reciprocal influence: A problem in treating sociometric data". *Canadian Journal of Experimental Psychology*. 3: 234.

Serre, D. 2010. *Matrices*. Springer-Verlag.

Shapley, L. S. 1953. "A Value for *n*-Person Games". In: *Contributions to the Theory of Games (AM-28), Volume II*. Ed. by H. W. Kuhn and A. W. Tucker. Princeton University Press.

Sinha, R. and R. Mihalcea. 2007. "Unsupervised graph-based word sense disambiguation using measures of word semantic similarity". In: *International Conference on Semantic Computing (ICSC 2007)*. IEEE. 363–369.

Skibski, O. and J. Sosnowska. 2018. "Axioms for distance-based centralities". In: *Proceedings of the AAAI Conference on Artificial Intelligence*, Vol. 32. No. 1.

Smirnova, E., K. Avrachenkov, and B. Trousse. 2010. "Using web graph structure for person name disambiguation.'. In: *Proceedings of CLEF*, Vol. 77. 80.

Solla Price, D. de. 1965. "Networks of Scientific Papers". *Science*. 149: 510–515.

Solla Price, D. de. 1976. "A general theory of bibliometric and other cumulative advantage processes". *Journal of the American society for Information science*. 27(5): 292–306.

Stankovic, L., D. Mandic, M. Dakovic, M. Brajovic, B. Scalzo, S. Li, and A. G. Constantinides. 2020. "Data Analytics on Graphs Part III: Machine Learning on Graphs, from Graph Topology to Applications". *Foundations and Trends in Machine Learning*. 13: 332–530.

Szczepański, P. L., T. P. Michalak, and T. Rahwan. 2016. "Efficient algorithms for game-theoretic betweenness centrality". *Artificial Intelligence*. 231: 39–63.

Traag, V. A., L. Waltman, and N. J. Van Eck. 2019. "From Louvain to Leiden: Guaranteeing well-connected communities". *Scientific Reports*. 9(1): 1–12.

Veremyev, A., O. A. Prokopyev, and E. L. Pasiliao. 2017. "Finding groups with maximum betweenness centrality". *Optimization Methods and Software*. 32(2): 369–399.

Vershynin, R. 2018. *High-dimensional Probability: An Introduction with Applications in Data Science*, Vol. 47. Cambridge University Press.

Vigna, S. 2016. "Spectral ranking". *Network Science*. 4(4): 433–445.

Volz, E. and D. D. Heckathorn. 2008. "Probability based estimation theory for respondent driven sampling". *Journal of Official Statistics*. 24(1): 79.

Von Luxburg, U. 2007. "A tutorial on spectral clustering". *Statistics and Computing*. 17(4): 395–416.

Wagner, D. and F. Wagner. 1993. "Between min cut and graph bisection". In: *International Symposium on Mathematical Foundations of Computer Science*. Springer. 744–750.

Wang, X. and I. Davidson. 2010. "Flexible constrained spectral clustering". In: *Proceedings of the 16th ACM SIGKDD*. 563–572.

Wang, X., B. Qian, and I. Davidson. 2014. "On constrained spectral clustering and its applications". *Data Mining and Knowledge Discovery*. 28(1): 1–30.

Was, T. and O. Skibski. 2018. "Axiomatization of the PageRank centrality". In: *Proceedings of IJCAI*. 3898–3904.

Wasserman, S. and K. Faust (1994. *Social Network Analysis: Methods and Applications*. Cambridge University Press.

Watts, D. J. 2000. *Small Worlds: The Dynamics of Networks between Order and Randomness*. Princeton University Press.

Watts, D. J. and S. H. Strogatz. 1998. "Collective dynamics of 'small-world' networks". *Nature*. 393(6684): 440–442.

Wei, T.-H. 1952. "Algebraic Foundations of Ranking Theory". *PhD thesis*. University of Cambridge.

West, J. D., M. C. Jensen, R. J. Dandrea, G. J. Gordon, and C. T. Bergstrom. 2013. "Author-level Eigenfactor metrics: Evaluating the influence of authors, institutions, and countries within the social science research network community". *Journal of the American Society for Information Science and Technology*. 64(4), 787–801.

White, S. and P. Smyth. 2003. "Algorithms for estimating relative importance in networks". In: *Proceedings of the 9-th ACM SIGKDD*. 266–275.

Wu, Z., S. Pan, F. Chen, G. Long, C. Zhang, and S. Y. Philip. 2020. "A comprehensive survey on graph neural networks". *IEEE Transactions on Neural Networks and Learning Systems*. 32(1): 4–24.

Xiao, H., K. Rasul, and R. Vollgraf. 2017. "Fashion-mnist: A novel image dataset for benchmarking machine learning algorithms". *arXiv preprint arXiv:1708.07747*.

Xu, K. S. and A. O. Hero. 2014. "Dynamic stochastic blockmodels for time-evolving social networks". *IEEE Journal of Selected Topics in Signal Processing*. 8(4): 552–562.

Xu, M., V. Jog, and P.-L. Loh. 2020. "Optimal rates for community estimation in the weighted stochastic block model". *Annals of Statistics*. 48(1): 183–204.

Yan, E. and Y. Ding. 2009. "Applying centrality measures to impact analysis: A coauthorship network analysis". *Journal of the American Society for Information Science and Technology*. 60(10), 2107–2118.

Yang, J. and J. Leskovec. 2015. "Defining and evaluating network communities based on ground-truth". *Knowledge and Information Systems*. 42(1): 181–213.

Yang, S., F. B. Keller, and L. Zheng. 2016. *Social Network Analysis: Methods and Examples*. SAGE Publications.

Yoshida, Y. 2014. "Almost linear-time algorithms for adaptive betweenness centrality using hypergraph sketches". In: *Proceedings of the 20th ACM SIGKDD*. 1416–1425.

Yu, Y., T. Wang, and R. J. Samworth. 2015. "A useful variant of the Davis–Kahan theorem for statisticians". *Biometrika*. 102(2): 315–323.

Yule, G. U. 1925. "A mathematical theory of evolution, based on the conclusions of Dr. JC Willis, FR S". *Philosophical Transactions of the Royal Society of London. Series B*. 213(402–410): 21–87.

Zachary, W. W. 1977. "An information flow model for conflict and fission in small groups". *Journal of Anthropological Research*. 33(4): 452–473.

Zhang, A. Y., H. H. Zhou, *et al.* 2016. "Minimax rates of community detection in stochastic block models". *Annals of Statistics*. 44(5): 2252–2280.

Zhang, L. and T. P. Peixoto. 2020. "Statistical inference of assortative community structures". *Physical Review Research*. 2(4): 043271.

Zhang, Y. and K. Rohe. 2018. "Understanding regularized spectral clustering via graph conductance". In: *Advances in Neural Information Processing Systems*. 10631–10640.

Zhou, D., O. Bousquet, T. N. Lal, J. Weston, and B. Schölkopf. 2004. "Learning with local and global consistency". In: *Advances in Neural Information Processing Systems*. 321–328.

Zhou, J., G. Cui, S. Hu, Z. Zhang, C. Yang, Z. Liu, L. Wang, C. Li, and M. Sun. 2020. "Graph neural networks: A review of methods and applications". *AI Open*. 1: 57–81.

Zhu, X. and Z. Ghahramani. 2002. 'Learning from labeled and unlabeled data with label propagation". *Technical Report CMU-CALD-02-107*. Carnegie Mellon University, Pittsburgh.

Zhu, X., Z. Ghahramani, and J. D. Lafferty. 2003. "Semi-supervised learning using Gaussian fields and harmonic functions". In: *Proceedings of the 20th International Conference on Machine Learning (ICML-03)*. 912–919.

Zhu, Z. A., S. Lattanzi, and V. Mirrokni. 2013. "A local algorithm for finding well-connected clusters". In: *International Conference on Machine Learning*. PMLR. 396–404.

Index

About the Authors

Konstantin Avrachenkov received his Master degree in Control Theory from St. Petersburg State Polytechnic University (1996), Ph.D. degree in Mathematics from University of South Australia (2000) and Habilitation from University of Nice Sophia Antipolis (2010). Currently, he is a Director of Research at Inria Sophia Antipolis, France. He is an associate editor of the International Journal of Performance Evaluation, Probability in the Engineering and Informational Sciences, ACM TOMPECS, Stochastic Models and IEEE Network Magazine. He has won 5 best paper awards. His main theoretical research interests are Markov chains, Markov decision processes, random graphs and singular perturbations. He applies these methodological tools to the modeling and control of networks, and to design data mining and machine learning algorithms.

Maximilien Dreveton received his Bachelor and Master degrees in the field of Physics from Ecole Normale Supérieure de Lyon, France, in 2013 and 2015. He obtained his Ph.D. in Computer Science from Inria Sophia Antipolis in 2022 and is currently a postdoctoral researcher at EPFL in Lausanne, Switzerland. His research interests include statistical analysis of random graphs, and more particularly community detection.

www.ingramcontent.com/pod-product-compliance
Lightning Source LLC
LaVergne TN
LVHW011509220926
848485LV00016B/75